Review of the Federal Strategy for Nanotechnology-Related Environmental, Health, and Safety Research

Committee for Review of the Federal Strategy to Address Environmental, Health, and Safety Research Needs for Engineered Nanoscale Materials

Board on Environmental Studies and Toxicology

Division on Earth and Life Studies

National Materials Advisory Board

Division on Engineering and Physical Sciences

NATIONAL RESEARCH COUNCIL
OF THE NATIONAL ACADEMIES

THE NATIONAL ACADEMIES PRESS
Washington, D.C.
www.nap.edu

THE NATIONAL ACADEMIES PRESS   500 Fifth Street, NW   Washington, DC 20001

NOTICE: The project that is the subject of this report was approved by the Governing Board of the National Research Council, whose members are drawn from the councils of the National Academy of Sciences, the National Academy of Engineering, and the Institute of Medicine. The members of the committee responsible for the report were chosen for their special competences and with regard for appropriate balance.

This project was supported by Contract No. NSFNNCO-073158 between the National Academy of Sciences and the National Science Foundation. Any opinions, findings, conclusions, or recommendations expressed in this publication are those of the author(s) and do not necessarily reflect the view of the organizations or agencies that provided support for this project.

International Standard Book Number-13: 978-0-309-11699-2
International Standard Book Number-10: 0-309-11699-6

Additional copies of this report are available from

The National Academies Press
500 Fifth Street, NW
Box 285
Washington, DC 20055

800-624-6242
202-334-3313 (in the Washington metropolitan area)
http://www.nap.edu

Printed in the United States of America.

THE NATIONAL ACADEMIES
Advisers to the Nation on Science, Engineering, and Medicine

The **National Academy of Sciences** is a private, nonprofit, self-perpetuating society of distinguished scholars engaged in scientific and engineering research, dedicated to the furtherance of science and technology and to their use for the general welfare. Upon the authority of the charter granted to it by the Congress in 1863, the Academy has a mandate that requires it to advise the federal government on scientific and technical matters. Dr. Ralph J. Cicerone is president of the National Academy of Sciences.

The **National Academy of Engineering** was established in 1964, under the charter of the National Academy of Sciences, as a parallel organization of outstanding engineers. It is autonomous in its administration and in the selection of its members, sharing with the National Academy of Sciences the responsibility for advising the federal government. The National Academy of Engineering also sponsors engineering programs aimed at meeting national needs, encourages education and research, and recognizes the superior achievements of engineers. Dr. Charles M. Vest is president of the National Academy of Engineering.

The **Institute of Medicine** was established in 1970 by the National Academy of Sciences to secure the services of eminent members of appropriate professions in the examination of policy matters pertaining to the health of the public. The Institute acts under the responsibility given to the National Academy of Sciences by its congressional charter to be an adviser to the federal government and, upon its own initiative, to identify issues of medical care, research, and education. Dr. Harvey V. Fineberg is president of the Institute of Medicine.

The **National Research Council** was organized by the National Academy of Sciences in 1916 to associate the broad community of science and technology with the Academy's purposes of furthering knowledge and advising the federal government. Functioning in accordance with general policies determined by the Academy, the Council has become the principal operating agency of both the National Academy of Sciences and the National Academy of Engineering in providing services to the government, the public, and the scientific and engineering communities. The Council is administered jointly by both Academies and the Institute of Medicine. Dr. Ralph J. Cicerone and Dr. Charles M. Vest are chair and vice chair, respectively, of the National Research Council.

www.national-academies.org

COMMITTEE FOR REVIEW OF THE FEDERAL STRATEGY TO ADDRESS
ENVIRONMENTAL, HEALTH, AND SAFETY RESEARCH NEEDS FOR
ENGINEERED NANOSCALE MATERIALS

Members

DAVID L. EATON *(Chair)*, University of Washington, Seattle
MARTIN A. PHILBERT *(Vice Chair)*, University of Michigan, Ann Arbor
GEORGE V. ALEXEEFF, California Environmental Protection Agency, Oakland
TINA BAHADORI, American Chemistry Council, Arlington, VA
JOHN M. BALBUS, Environmental Defense Fund, Washington, DC
MOUNGI G. BAWENDI, Massachusetts Institute of Technology, Cambridge
PRATIM BISWAS, Washington University at St. Louis, St. Louis, MO
VICKI L. COLVIN, Rice University, Houston, TX
STEPHEN J. KLAINE, Clemson University, Pendleton, SC
ANDREW D. MAYNARD, Woodrow Wilson International Center for Scholars,
    Washington, DC
NANCY A. MONTEIRO-RIVIERE, North Carolina State University, Raleigh
GUNTER OBERDÖRSTER, University of Rochester School of Medicine and
    Dentistry, Rochester, NY
MARK A. RATNER, Northwestern University, Evanston, IL
JUSTIN G. TEEGUARDEN, Pacific Northwest National Laboratory,
    Richland, WA
MARK R. WIESNER, Duke University, Durham, NC

Staff

EILEEN N. ABT, Project Director
MICHAEL MOLONEY, Senior Program Officer
NORMAN GROSSBLATT, Senior Editor
HEIDI MURRAY-SMITH, Research Associate
MIRSADA KARALIC-LONCAREVIC, Manager, Technical Information Center
PANOLA GOLSON, Senior Program Assistant

Sponsor

NATIONAL NANOTECHNOLOGY COORDINATION OFFICE

Acute Exposure Guideline Levels for Selected Airborne Chemicals (seven
    volumes, 2000-2008)
Toxicological Effects of Methylmercury (2000)
Strengthening Science at the U.S. Environmental Protection Agency (2000)
Scientific Frontiers in Developmental Toxicology and Risk Assessment (2000)
Ecological Indicators for the Nation (2000)
Waste Incineration and Public Health (2000)
Hormonally Active Agents in the Environment (1999)
Research Priorities for Airborne Particulate Matter (four volumes, 1998-2004)
The National Research Council's Committee on Toxicology: The First 50
    Years (1997)
Carcinogens and Anticarcinogens in the Human Diet (1996)
Upstream: Salmon and Society in the Pacific Northwest (1996)
Science and the Endangered Species Act (1995)
Wetlands: Characteristics and Boundaries (1995)
Biologic Markers (five volumes, 1989-1995)
Science and Judgment in Risk Assessment (1994)
Pesticides in the Diets of Infants and Children (1993)
Dolphins and the Tuna Industry (1992)
Science and the National Parks (1992)
Human Exposure Assessment for Airborne Pollutants (1991)
Rethinking the Ozone Problem in Urban and Regional Air Pollution (1991)
Decline of the Sea Turtles (1990)

Copies of these reports may be ordered from the National Academies Press
(800) 624-6242 or (202) 334-3313
www.nap.edu

Assessment of Practicality of Pulsed Fast Neutron Analysis for Aviation
     Security (2002)
Workshop on Technical Strategies for Adoption of Commercial Materials and
     Processing Standards in Defense Procurement (2002)
Summary of the Workshop on Structural Nanomaterials (2001)
Materials Research to Meet 21st Century Defense Needs: Interim Report (2001)
Small Business Innovation Research to Support Aging Aircraft: Priority
     Technical Areas and Process Improvements (2001)
Materials Technologies for the Process Industries of the Future: Management
     Strategies and Research Opportunities (2000)
Uninhabited Air Vehicles: Enabling Science for Military Systems (2000)
Summary Record of the Workshop on Polymer Materials Research – August
     30-31, 1999 (2000)

Copies of these reports may be ordered from the National Academies Press
(800) 624-6242 or (202) 334-3313
www.nap.edu

Preface

Nanotechnology relies on the ability to engineer, manipulate, and manufacture materials at the nanoscale. Nanotechnology is already enabling the development of an industry that produces and uses engineered nanomaterials in a wide variety of industrial and consumer products. The increasing use of nanomaterials in industrial and consumer products will result in greater exposure of workers and the general public to engineered nanoscale materials.

The U.S. National Nanotechnology Initiative (NNI) is the central locus for the coordination of federal agency investments in nanoscale research and development. In 2007, the National Nanotechnology Coordination Office, which oversees the operation of NNI, asked the National Research Council to review its publication *Strategy for Nanotechnology-Related Environmental, Health, and Safety Research*. The National Research Council's Board on Environmental Studies and Toxicology and National Materials Advisory Board convened the Committee for Review of the Federal Strategy to Address Environmental, Health, and Safety Research Needs for Engineered Nanoscale Materials, which produced this report. The committee was composed of members with expertise in nanotechnology, nanomaterials, metrology, toxicology, risk assessment, exposure assessment, ecotoxicology, occupational and public health, and risk management.

The committee was asked to conduct a scientific and technical review of the federal strategy. The committee considered the elements of an effective nanotechnology risk-research strategy, evaluated whether the federal strategy has these elements, and assessed how the research identified in the strategy will support risk-assessment and risk-management needs. To assist its task, the committee held two workshops at which it heard from representatives of NNI agencies, policy experts from the European Commission, and such stakeholders as manufacturing industry, nongovernment organizations, and the insurance sector.

This report has been reviewed in draft form by persons chosen for their diverse perspectives and technical expertise in accordance with procedures approved by the National Research Council's Report Review Committee. The purpose of this independent review is to provide candid and critical comments that will assist the institution in making its published report as sound as possible and to ensure that the report meets institutional standards of objectivity, evidence, and responsiveness to the study charge. The review comments and draft manuscript remain confidential to protect the integrity of the deliberative process. We wish to thank the following for their review of this report: David E. Aspnes, North Carolina State University; Chris G. Whipple, ENVIRON International Corporation; Richard A. Denison, Environmental Defense Fund; William H. Farland, Colorado State University; Richard A.L. Jones, University of Sheffield; Gregory V. Lowry, Carnegie Mellon University; David Y. Pui, University of Minnesota; Ronald F. Turco, Purdue University; Mark J. Utell, University of Rochester School of Medicine and Dentistry; David B. Warheit, DuPont Haskell Laboratory.

Although the reviewers listed above have provided many constructive comments and suggestions, they were not asked to endorse the conclusions or recommendations, nor did they see the final draft of the report before its release. The review of the report was overseen by the review coordinator, Richard Schlesinger, Pace University, and the review monitor, Elsa Garmire, Dartmouth College. Appointed by the National Research Council, they were responsible for making certain that an independent examination of the report was carried out in accordance with institutional procedures and that all review comments were carefully considered. Responsibility for the final content of the report rests entirely with the committee and the institution.

The committee gratefully acknowledges the following for their presentations: Pilar Aguar, European Commission; Norris Alderson, U.S. Food and Drug Administration; Carolyn Cairns, Consumers Union; Richard Canady, U.S. Food and Drug Administration; Altaf Carim, U.S. Department of Energy; Thomas Epprecht, Swiss Re; William Gulledge, American Chemistry Council; Michael Holman, Lux Research; William Kojola, AFL-CIO; Philippe Martin, European Commission; Terry Medley, DuPont; Jeffrey Morris, U.S. Environmental Protection Agency; Vladimir Murashov, National Institute for Occupational Safety and Health; Dianne Poster, National Institute of Standards and Technology; William Rees, U.S. Department of Defense; Mihail Roco, National Science Foundation; Jennifer Sass, National Resources Defense Council; Phillip Sayre, U.S. Environmental Protection Agency; Paul Schulte, National Institute for Occupational Safety and Health; Clayton Teague, National Nanotechnology Coordination Office; and Sally Tinkle, National Institute of Environmental Health Sciences.

The committee is also grateful for the assistance of the National Research Council staff in preparing this report. Staff members who contributed to the effort are Eileen Abt, project director; Michael Moloney, senior program officer; James Reisa, director of the Board on Environmental Studies and Toxicology; Heidi Murray-Smith, research associate; Norman Grossblatt, senior editor; Mirsada Karalic-Loncarevic, manager, technical information center; and Panola Golson, senior program assistant.

We would especially like to thank the committee members for their efforts throughout the development of this report.

David L. Eaton, *Chair*
Martin A. Philbert, *Vice Chair*
Committee for Review of the Federal Strategy to
Address Environmental, Health, and Safety Research
Needs for Engineered Nanoscale Materials

Abbreviations

ADME	absorption, distribution, metabolism, elimination
AEC	Atomic Energy Commission
CSREES	Cooperative State Research, Education, and Extension Service
CST	UK Council for Science and Technology
DHS	Department of Homeland Security
DHHS	Department of Health and Human Services
DOC	Department of Commerce
DOD	Department of Defense
DOE	Department of Energy
DOJ	Department of Justice
DOT	Department of Transportation
EC	European Commission
EHS	environmental, health, and safety
EPA	U.S. Environmental Protection Agency
EU	European Union
FDA	Food and Drug Administration
FHWA	Federal Highway Administration
FS	Forest Service
FY	fiscal year
GIN	Global Issues in Nanotechnology Working Group
ICON	International Council on Nanotechnology
IWGN	Interagency Working Group on Nanotechnology
NASA	National Aeronautics and Space Administration
NEHI	Nanotechnology Environmental Health Implications
NIH	National Institutes of Health
NILI	Nanomanufacturing Industry Liaison and Innovation Working Group
NIOSH	National Institute for Occupational Safety and Health
NIST	National Institute of Standards and Technology
NNCO	National Nanotechnology Coordination Office
NNI	U.S. National Nanotechnology Initiative
NORA	National Occupational Research Agenda

NPEC	National Public Engagement and Communications Working Group
NRC	National Research Council
NRC	Nuclear Regulatory Commission
NSET	Nanoscale Science, Engineering, and Technology subcommittee
NSF	National Science Foundation
NSTC	National Science and Technology Council
OECD	Organization for Economic Co-operation and Development
OMB	Office of Management and Budget
OSTP	Office of Science and Technology Policy
PART	Program Assessment Rating Tool
PCA	program component area
PCAST	President's Council of Advisors on Science and Technology
QSAR	quantitative structure–activity relationship
R&D	research and development
SCENIHR	Scientific Committee on Emerging and Newly-Identified Health Risks
USDA	U.S. Department of Agriculture
VOI	Value of Information
WPMN	Working Party on Manufactured Nanomaterials

Contents

BOXES, FIGURES, AND TABLES

BOXES

FIGURES

TABLES

Review of the Federal Strategy for Nanotechnology-Related Environmental, Health, and Safety Research

Summary

The field of nanotechnology relies on the ability to engineer, manipulate, and manufacture materials at the nanoscale.[1] Nanotechnology is already enabling the development of an industry that produces and uses engineered nanomaterials in a wide variety of industrial and consumer products, such as targeted drugs, video displays, remediation of groundwater contaminants, high performance batteries, dirt-repelling coatings on building surfaces and clothing, high-end sporting goods, and skin-care products. Over the next five to ten years, increasingly widespread use of complex engineered nanomaterials is anticipated in such products as medical treatments, super-strong lightweight materials, food additives, and advanced electronics. The increasing use of engineered nanoscale materials in industrial and consumer products will result in greater exposure of workers and the general public to these materials. Responsible development of nanotechnology implies a commitment to develop and to use these materials to meet human and societal needs while making every reasonable effort to anticipate and mitigate adverse effects and unintended consequences.

The U.S. National Nanotechnology Initiative (NNI) is the government's central locus for the coordination of federal agency investments in nanoscale research and development. NNI is responsible for supporting the missions of its member research and regulatory agencies; ensuring U.S. leadership in nanoscale science, engineering, and technology; and contributing to the nation's economic competitiveness. Within NNI, the Nanotechnology Environmental Health Implications (NEHI) Working Group provides a forum for the NNI agencies to coordinate their activities related to understanding the potential risks posed by nanotechnology to protect public health and the environment.[2] The NEHI's co-

---

[1]Nanoscale refers to materials on the order of one billionth of a meter.

[2]Current members of NEHI consists of officials from the Consumer Product Safety Commission, Cooperative State Research, Education, and Extension Service, Department

3

ordination efforts have produced a series of documents that identify environmental, health, and safety (EHS) research needs related to nanomaterials (NEHI 2006, 2007, 2008).[3]

In 2007, the National Nanotechnology Coordination Office, which oversees the day-to-day operations of the NNI, asked the National Research Council to review independently its *Strategy for Nanotechnology-Related Environmental, Health, and Safety Research* (NEHI 2008). In response, the National Research Council's Board on Environmental Studies and Toxicology and National Materials Advisory Board oversaw the appointment of the Committee for Review of the Federal Strategy to Address Environmental, Health, and Safety Research Needs for Engineered Nanoscale Materials, which produced this report. The committee was charged to conduct a scientific and technical review of the federal strategy and to comment in general terms on how the strategy develops information needed to support EHS risk-assessment and risk-management needs with respect to nanomaterials.

Assisted by information-gathering sessions that included representatives from NNI agencies, policy experts from the European Commission, and such stakeholders as manufacturing industry, nongovernment organizations, and the insurance sector, the committee evaluated the federal strategy, asking such questions as the following:

• What are the elements of an effective nanotechnology risk-research strategy?
• Does the federal strategy have those elements?
• With respect to the federal strategy, have the appropriate research needs been identified, are the gap analysis and the selection of priorities among research needs complete, and does the research identified support risk-assessment and risk-management needs?

---

of Defense, Department of Energy, Department of State, Department of Transportation, Environmental Protection Agency, Food and Drug Administration, International Trade Commission, National Aeronautics and Space Administration, National Institute for Occupational Safety and Health, National Institutes of Health, National Institute of Standards and Technology, National Science Foundation, Occupational Safety and Health Administration, Office of Science and Technology Policy, Office of Management and Budget, and U.S. Geological Survey.

[3]NEHI (Nanotechnology Environmental Health Implications Working Group). 2006. Environmental, Health, and Safety Research Needs for Engineered Nanoscale Materials. Arlington, VA: National Nanotechnology Coordination Office; NEHI (Nanotechnology Environmental Health Implications Working Group). 2007. Prioritization of Environmental, Health, and Safety Research Needs for Engineered Nanoscale Materials: An Interim Document for Public Comment. Arlington, VA: National Nanotechnology Coordination Office; NEHI (Nanotechnology Environmental Health Implications Working Group). 2008. Strategy for Nanotechnology-Related Environmental, Health, and Safety Research. Arlington, VA: National Nanotechnology Coordination Office.

WHAT ARE THE ELEMENTS OF AN EFFECTIVE
NANOTECHNOLOGY RISK-RESEARCH STRATEGY?

Strategies for conducting scientific research are particularly important when resources are limited and there is a need to ensure that relevant information is being generated as efficiently and cost-effectively as possible. A strategy generally defines a set of goals, often in the context of an overarching vision; a plan of action for achieving the goals; and milestones to indicate when the goals are expected to be achieved. Because scientific research is often open-ended and serendipitous, formulating goals can be difficult.

One specific type of research strategy—a strategy for risk research—addresses challenges of broad societal significance: the reduction or prevention of harm to humans and the environment. Because of their potential influence on public-health and environmental policy and actions, it is critical that risk-research strategies be developed and implemented effectively and in a timely manner. And like any other risk-research strategy, one focused on nanotechnology-related risk research needs to be proactive—identifying possible risks and ways to mitigate risks before the technology has widespread commercial presence. It has to address nanotechnology-based products that are beginning to enter commerce as well as those under development. But it also needs to lay the scientific groundwork for addressing materials and products that potentially will arise out of new research, new tools, and cross-fertilization between distinct fields of science and technology. Therefore, a nanotechnology-related risk-research strategy must rely on both targeted research, which addresses questions that are critical for ensuring the safety of nanomaterials and products that contain them, and exploratory research, which generates new knowledge that will inform future goals and research directions.

In conducting this study, the committee identified nine elements that are integral to any effective risk-research strategy and that informed its evaluation of the 2008 NNI document:

- *Vision, or statement of purpose.* What is the ultimate purpose of conducting research on potential risks associated with nanotechnology?
- *Goals.* What specific research goals need to be achieved to guide the development and implementation of nanotechnologies that are as safe as possible?
- *Evaluation of the state of science.* What is known about the potential for the products of nanotechnology to cause harm and about how possible risks might be managed?
- *Road map.* What is the plan of action to achieve the stated research goals?
- *Evaluation.* How will research progress be measured, and who will be responsible for measuring it? Are there measurable milestones that can be evaluated against a clear timeline?

- *Review.* How will the strategy be revised in light of new findings, to ensure that it remains responsive to the overarching vision and goals?
- *Resources.* Are there sufficient resources to achieve the stated goals? If not, what are the plans to obtain new resources or to leverage other initiatives to achieve the goals?
- *Mechanisms.* What are the most effective approaches to achieving the stated goals?
- *Accountability.* How will stakeholders participate in the process of developing and evaluating a research strategy? Who will be accountable for progress toward stated goals?

DOES THE FEDERAL STRATEGY HAVE THOSE ELEMENTS?

On the basis of the information gathered at its public meetings and the professional expertise and experience of its members, the committee determined that the process of composing the government's 2008 NNI document provided a unique and useful opportunity for coordination, planning, and consensus-building among NEHI-member federal agencies. The strategy demonstrates how the NNI and the agencies have effectively worked together to coordinate their funding and their assessment of EHS aspects of nanotechnology.

However, NNI (NEHI 2008) does not have the essential elements of a research strategy—it does not present a vision, contain a clear set of goals, have a plan of action for how the goals are to be achieved, or describe mechanisms to review and evaluate funded research and assess whether progress has been achieved in the context of what we know about the potential EHS risks posed by nanotechnology.

The NNI document contains various statements of purpose, but it does not provide a clear vision as to where our understanding of the EHS implications of nanotechnology should be in 5 or 10 years. It states that "the NEHI Working Group developed this nanotechnology-related EHS research strategy to accelerate progress in research to protect public health and the environment, and to fill gaps in, and—with the growing level of effort worldwide—to avoid unnecessary duplication of, such research" (NEHI 2008, p. 1). That statement of purpose is adequate for an open-ended research program with no definite objectives, but it falls short of ensuring that the results of strategic research are useful and applicable to decision-making that will reduce the potential environmental and health effects of nanotechnology.

The strategy document does not present goals for research to help ensure that the development and implementation of nanotechnology is as safe as practicable or a road map to ensure that these research goals are achieved. Although the document identifies five "research needs" for each of five research categories—"Instrumentation, Metrology, and Analytical Methods," "Nanomaterials and Human Health," "Nanomaterials and the Environment," "Human and Environmental Exposure Assessment," and "Risk Management Methods"—the needs are not articulated as clear goals that should be attained. A key element of any strategy is to identify goals and measures of progress or success before assessing what is being done. That allows a clear assessment of the value of current activities. Such an approach enables development of an action plan to leverage other efforts and address research deficiencies in a way that is transparent and measurable. Because the NNI document does not establish goals and a plan of action, there is no element of accountability, and questions are never raised as to what other research activities are needed.

The NNI document does not provide an evaluation of the state of science in each of the five research categories; rather, the research needs are evaluated against research projects that were funded in FY 2006 (see Appendix A of NEHI [2008]) to provide a "snapshot" of research activities. The 2008 NNI document uses the FY 2006 data to assess the extent to which federally funded EHS research related to nanomaterials is supporting selected research priorities and to conduct its gap analysis of the NNI research portfolio. The committee concludes that how the FY 2006 data were used in the analysis is probably the greatest deficiency in the 2008 document, inasmuch as it is the foundation of the document's evaluation of the strengths, weaknesses, and gaps in currently funded federal research. This is problematic because most of the listed FY 2006 research projects were focused on understanding fundamentals of nanoscience that are not explicitly associated with risk or the development of nanotechnology applications.[4] In addition, there is no clear statement of how the FY 2006 research projects would address the identified research needs and inform an understanding of potential human health and environmental risks posed by engineered nanoscale materials.

The 2008 document does provide some information on time frame and sequencing for achieving the research needs (see Figures 3, 5, 7, 9, and 11 of NNI [NEHI 2008]) but with little justification.

The NNI strategy does not identify resources necessary to address questions concerning EHS research needs for nanomaterials. Although the detailed analysis of nanotechnology-related EHS expenditures in FY 2006 provides information about what was spent during that year, there is no assessment of whether the aggregate level of spending was adequate to address EHS research needs or whether the resource expenditures by the agencies were appropriate to address EHS research needs based on their missions. An appropriate research strategy would quantify the resources needed to address research priorities and describe where the resources would come from.

---

[4]The 246 FY 2006 research projects listed in NNI (NEHI 2008) include additional research on instrumentation and metrology research and on medical-application-oriented research that is not captured in the list of 130 EHS research projects in the annual supplement to the president's budget. The committee's own assessment of the number of FY2006 research projects that are relevant to understanding risk of nanomaterials is discussed in Chapter 4.

Although lead agencies (for example, NIH, NIST, EPA, FDA, and NIOSH) are given roles for overseeing federal nanotechnology research, there is no accountability, that is, there is no single organization or person that will be held accountable for whether the government's overall strategy delivers results. Accountability requires specific quantifiable objectives so that one can determine whether adequate progress is being made. The 2008 NNI document does not adequately incorporate input from other stakeholders, such as industries that produce nanomaterials and end users of nanomaterials; environmental and consumer advocacy groups; foreign interests, including substantial efforts of other countries; and local and state governments. The committee recognizes that the 2006 and 2007 NNI reports have undergone public comment, but public comment is not the same as engaging stakeholders in the process.

Without adequate input from external stakeholders, it is not possible for government agencies to develop an effective research strategy to underpin the emergence of safe nanotechnologies. Federal agencies may have a vested interest in justifying the value of current efforts rather than critically assessing what is being done and how deficiencies might be addressed. For example, when developing their own research strategies, agencies tend to ask, What research can we do within our existing capabilities?, rather than the more appropriate question, What research should we be doing?

REVIEW OF PRIORITY RESEARCH TOPICS, RESEARCH NEEDS, AND GAP ANALYSIS

The committee reviewed the specific research categories and their designated research needs as described in the 2008 NNI document (Section II) and considered the following questions: Were the appropriate research needs identified? Were the gap analysis and priority sequencing of research needs complete? Does the identified research support risk-assessment and risk-management needs?

The NNI's five topical categories each address research that is important for EHS risk assessment and risk management, and collectively they cover the necessary broad research topics. The listed research needs in the five categories are similarly valuable but incomplete, in some cases missing elements crucial for progress in understanding the EHS implications of nanomaterials. For example, the subject of environmental exposure received insufficient emphasis in the exposure-assessment discussion, and characterization of chemical and biologic reactivity of nanoparticles was not included as a research need. That appears to have resulted from an effort to place research needs into one of the five "silo" categories with little discussion of the interrelationships and interconnections among categories.

The committee notes examples of other research needs that it judged to be insufficiently addressed in the document. For "Nanomaterials and Human Health," a more comprehensive analysis and evaluation of absorption, distribution, metabolism, elimination, and toxicity of engineered nanomaterials at realistic exposure levels is needed. For "Human and Environmental Exposure," exposures throughout the life cycle of nanomaterials was not sufficiently introduced or adequately integrated into this section, although a discussion was contained within "Risk Management Methods."

The NNI's gap analysis is not accurate in that the relevance of FY 2006 research projects to the research needs is generally overstated. The 2008 document consistently—in every research category—appears to assume that funded projects with only distant links to a research question were meeting that research need. In the "Nanomaterials and Human Health" category, more than 50% of the inventoried projects describe research directly relevant to developing therapeutic strategies aimed at cancer and other ailments rather than any of the research needs listed as relevant to potential EHS risks posed by nanomaterials. The committee acknowledges the value of therapeutic research but believes that it is not directly relevant to understanding potential risks associated with nanomaterials that are important in occupational, environmental, and ecologic exposure scenarios. In the category of risk-management methods, there is no coverage of management of environmental and consumer risks, including specific potential exposure scenarios, such as accidents and spills, environmental discharges, and exposure through consumer products. Uniformly, the committee agreed that many of the 246 research projects listed in Appendix A are of high scientific value, but the vast majority are of little or no direct value in reducing the uncertainty faced by stakeholders making decisions about nanotechnology and its risk-management practices. The 2008 document substantially overestimates the general nanotechnology-related research activity in environmental, health, and safety research.

In many cases, the committee concluded that the sequencing of research needs was generally appropriate but not adequately justified. In a number of cases the committee questioned the rationale for a sequence. For example, in the "Instrumentation, Metrology, and Analytical Methods" category, why put the development of materials to support exposure assessment before materials to support toxicology studies? Why delay research into alternative surface-area measurement methods for 10 years if it is identified as a critical research subject? In the "Nanomaterials and the Environment" category, the committee questioned whether resources could be used more efficiently through the characterization of exposure and transformation processes prior to characterization of organisms as well as higher-level ecosystem effects.

Although many of the NNI's identified research needs support risk-assessment and risk-management needs, the committee concluded that failure to identify important research needs, the lack of rationale for and discussion of research priorities, and the flaws in the gap analysis undermine the ability to ensure that currently funded research adequately supports EHS risk-assessment and risk-management needs and provides critical data for the federal agencies.

CONCLUSIONS AND RECOMMENDATIONS

The NNI's 2008 *Strategy for Nanotechnology-Related Environmental, Health, and Safety Research* **could be an effective tool for communicating the breadth of federally supported research associated with developing a more complete understanding of the environmental, health, and safety implications of nanotechnology. It is the result of considerable collaboration and coordination among 18 federal agencies and is likely to eliminate unnecessary duplication of their research efforts. However, the document does not describe a strategy for nano-risk research. It lacks input from a diverse stakeholder group, and it lacks essential elements, such as a vision and a clear set of objectives, a comprehensive assessment of the state of the science, a plan or road map that describes how research progress will be measured, and the estimated resources required to conduct such research.**

There remains an urgent need for the nation to build on the current research base related to the EHS implications of nanotechnology— including the federally supported research as described in the 2008 NNI document—by developing a national strategic plan for nanotechnology-related environmental, health, and safety research.

A national strategic plan for nanotechnology-related EHS research would identify research needs clearly and estimate the financial and technical resources required to address identified research gaps. It would also provide specific, measurable objectives and a timeline for meeting them. The national strategic plan, unlike the 2008 NNI document, would consider the untapped knowledge of and input from nongovernment researchers and academics, who can contribute to understanding the potential EHS implications of nanotechnology.

Reducing the burden of uncertainty through targeted, effective research that identifies and eliminates potential environmental and health hazards of engineered nanoscale materials should have high priority for the nation. An effective national EHS strategic research plan is essential to the successful development of and public acceptance of nanotechnology-enabled products. This strategy should be informed by value-of-information thinking to determine the research that is needed to reduce the current uncertainties with respect to the potential health and environmental effects of nanomaterials. A national strategic plan would need to address nanotechnology-based products that are entering commerce as well as nanotechnologies that are under development. It would provide a path to developing the scientific knowledge to support nanotechnology-related EHS risk-based decision-making.

The committee concludes that a truly national strategy cannot be developed within the limitations of the scope of research under the umbrella of the NNI. Although the 2008 NNI document potentially represents excellent input into the national strategic plan, the NNI can produce only a strategy that is the sum of the individual agency strategies and priorities. The structure of the NNI makes the development of a visionary and authoritative research strategy extraordinarily difficult. Because the NNI is not a research funding program but rather a coordination mechanism, comprising the activities of 25 federal agencies, it has no central authority to make budgetary or funding decisions, and it relies on the budgets of its member agencies to gather resources or influence the shape of the overall federal nanotechnology-related EHS research activity. Because the NNI is responsible for ensuring U.S. competitiveness through the rapid development of a robust research and development program in nanotechnology while ensuring the safe and responsible development of nanotechnology, it may be perceived as having a conflict of interest. But the conflict is a false dichotomy. Strategic research on potential risks posed by nanotechnology should be an integral and fundamental part of the sustainable development of nanotechnology. Nonetheless, a clear separation of accountability for development of applications and assessment of potential implications of nanotechnology would help to ensure that the public-health mission has appropriate priority.

The committee is concerned that the actual amount of federal funding specifically addressing the EHS risks posed by nanotechnology is far less than portrayed in the NNI document and may be inadequate. The committee concludes that if no new resources are provided and the current levels of agency funding continue, the research that is generated cannot adequately evaluate the potential health and environmental risks and effects associated with engineered nanomaterials to address the uncertainties in current understanding. Such an evaluation is critical for ensuring that the future of nanotechnology is not burdened by uncertainties and innuendo about potential adverse health and environmental effects. Those concerns have been voiced recently by the nanotechnology industry and various environmental and public-health interest groups.

Having reviewed the 2008 NNI strategy document and discussed what is needed for a path forward, the committee presents the following recommendations:

A robust national strategic plan is needed for nanotechnology-related environmental, health, and safety research that builds on the five categories of research needs identified in the 2008 NNI document. The development of the plan should include input from a broad set of stakeholders across the research community and other interested parties in government, nongovernment, and industrial groups. The strategy should focus on research to support risk assessment and management, should include value-of-information considerations, and should identify

> - **Specific research needs for the future in such topics as potential exposures to engineered nanomaterials, toxicity, toxicokinetics, environmental fate, and standardization of testing.**
> - **The current state of knowledge in each specific area.**

- **The gap between the knowledge at hand and the knowledge needed.**
- **Research priorities for understanding life-cycle risks to humans and the environment.**
- **The estimated resources that would be needed to address the gap over a specified time frame.**

As part of a broader strategic plan, NNI should continue to foster the successful interagency coordination effort that led to its 2008 document with the aim of ensuring that the federal plan is an integral part of the broader national strategic plan for investments in nanotechnology-related environmental, health, and safety research. In doing so, it will need a more robust gap analysis. The federal plan should identify milestones and mechanisms to ascertain progress and identify investment strategies for each agency. Such a federal plan could feed into a national strategic plan but would not itself be a broad, multistakeholder national strategic plan. Development of a national strategic plan should begin immediately and not await further refinement of the current federal strategy.

CONCLUDING REMARKS

A robust national strategic plan for addressing nanotechnology-related EHS risks will need to focus on promoting research that can assist all stakeholders, including federal agencies, in planning, controlling, and optimizing the use of engineered nanomaterials while minimizing EHS effects of concern to society. Such a plan will ensure the timely development of engineered nanoscale materials that will bring about great improvements in the nation's health, its environmental quality, its economy, and its security.

1

The National Nanotechnology Initiative and the Genesis of the Environmental, Health, and Safety Strategy

Nanotechnology consists of several enabling technologies that take advantage of unique properties of extremely tiny structures in applications ranging from medicine to electronics to material science. Research in nanotechnology is based on understanding the physical and chemical properties of materials at the level of molecules or complexes of molecules, or atomic clusters with the goal to be able to manipulate those properties. Nanotechnology is not simply about small particles, materials, or products and is defined by the federal government as including the following three factors (NSET 2008a):

- Research and technology development at the atomic, molecular, or macromolecular levels on a length scale of about 1-100 nm (a nanometer is one-billionth of a meter—too small to be seen with a conventional optical microscope).
- Creation and use of structures, devices, and systems that have novel properties and functions because they are small or of intermediate size, specifically, at the level of atoms and molecules.
- Ability to control or manipulate materials on an atomic scale.

In the middle 1990s, as better methods for the characterization, processing, and manipulation of matter on the nanoscale were being developed in research programs supported by federal science and technology agencies, the agencies began holding informal discussions on a common vision for nanotechnology. The interagency dialogue culminated in the establishment in 2000 of the National Nanotechnology Initiative (NNI)—Box 1-1 details some of the history of the establishment of the initiative. The NNI serves strictly as a coordination mechanism for government agencies that support nanoscale research, such as the Department of Energy and the National Science Foundation, or that have a stake

BOX 1-1 A Brief History of the National Nanotechnology Initiative

In September 1998, an interagency dialogue on nanotechnology was formalized as the Interagency Working Group on Nanotechnology (IWGN). Established under the National Science and Technology Council (NSTC) of the Office of Science and Technology Policy, the IWGN developed a number of reports on a long-term vision for nanoscale research and development (R&D), on international benchmarking of nanotechnology, and on U.S. government investment in nanotechnology R&D (Siegel et al. 1999; Roco et al. 2001). In March 1999, IWGN representatives proposed a nanotechnology initiative with a budget of a half-billion dollars for FY 2001 (Roco 2004). In November 2000, the National Nanotechnology Initiative (NNI) was formally established, and preparations were begun for a coordinated federal investment in nanoscale R&D.

In August 2000, as the NNI proposal matured, the NSTC established the Nanoscale Science, Engineering and Technology (NSET) Subcommittee to replace the IWGN. The NSET Subcommittee was tasked with implementing the NNI by coordinating with federal agencies and R&D programs. Beginning with eight agencies in 2001, the subcommittee now comprises representatives of over 25 federal departments and agencies and officials of the White House Office of Science and Technology Policy and the White House Office of Management and Budget.

In January 2001, the National Nanotechnology Coordination Office (NNCO) was established to provide daily technical and administrative support to the NSET Subcommittee and to assist in multiagency planning and the preparation of budgets and program-assessment documents. The NNCO was also tasked with assisting the NSET Subcommittee with the collection and dissemination of information on industry, state, and international nanoscale science and technology research, development, and commercialization activities (NRC 2002). The NNCO provides technical guidance and administrative support, organizes monthly NSET Subcommittee meetings, conducts workshops, and prepares information and reports, serving as a point of contact and helping to facilitate communication.

in the outcomes of nanoscale research, such as the Food and Drug Administration (FDA) and the Department of Justice. Under the broad umbrella of the initiative, each participating agency invests in projects and programs in support of its own mission. The NNI consists of individual and cooperative nanotechnology-related activities of 25 federal agencies with a wide array of research and regulatory responsibilities. The NNI itself does not fund research, and its budget is equal to the sum of the amounts at which member agencies fund their individual or joint nanotechnology-related programs and projects. Therefore, the NNI has no authority to make budgetary or funding decisions; it relies on the budgets of its member agencies. The goals of the NNI are as follows (NSET 2008a):

- Advance a world-class nanotechnology research and development program.
- Foster the transfer of new technologies into products for commercial and public benefit.
- Develop and sustain educational resources, a skilled workforce, and the supporting infrastructure and tools to advance nanotechnology.
- Support responsible development of nanotechnology.

The NNI's primary coordination mechanism is the National Science and Technology Council (NSTC) Nanoscale Science, Engineering, and Technology (NSET) Subcommittee (NSET 2008a). Through the operation of the NSET Subcommittee and subordinate structures of the NNI, the initiative addresses the general goals of supporting the missions of the participating agencies; ensuring continuing leadership by the United States in nanoscale science, engineering, and technology; and contributing to the nation's economic competitiveness.

In 2003, the 21st Century Nanotechnology Research and Development Act (Public Law 108-153) was signed into law. The legislation established the NNI's operating structures and required that the president establish or designate an advisory panel with a membership qualified to provide advice and information on nanotechnology research, development, demonstrations, education, technology transfer, commercial applications, and societal and ethical concerns.[1] The President's Council of Advisors on Science and Technology (PCAST) was assigned by the president to play such a role. Figure 1-1 shows the current organizational structure of the NNI.

Thirteen NNI-participating agencies currently report investments in nanotechnology: the Department of Agriculture (USDA) (including the Forest Service [FS] and the Cooperative State Research, Education, and Extension Service [CSREES]), Department of Defense (DOD), Department of Energy (DOE), Department of Homeland Security (DHS), Department of Justice (DOJ), Department of Transportation (DOT), Environmental Protection Agency (EPA), National Aeronautics and Space Administration (NASA), National Institute for Occupational Safety and Health (NIOSH), National Institute of Standards and Technology (NIST), National Institutes of Health (NIH), and National Science Foundation (NSF). In FY 2007, the total investment by those agencies in NNI-related research was about $1.425 billion; DOD, DOE, NIH, NIST, and NSF contributed over 80% of the total NNI budget. The president's research and development (R&D) budget request for the NNI for FY 2009 was $1.527 billion.

Released in December 2007, the updated NNI strategic plan (NSET 2007a) looks 5-10 years ahead to outline a vision of the NNI as working for a "future in which the ability to understand and control matter at the nanoscale

---

[1]Such a panel had been called for in Small Wonders, Endless Frontiers: A Review of the National Nanotechnology Initiative (NRC 2002).

leads to a revolution in technology and industry that benefits society" (NSET 2007a, p. 3).

The strategic plan outlines program component areas (PCAs)[2] that were developed as a means of categorizing and describing the many investments in nanotechnology R&D by the federal agencies that support research. Table 1-1 shows the FY 2008 estimated agency expenditures for the PCAs among the NNI agencies. The committee notes that there may be additional nanotechnology research being performed by some agencies that is not reported in the table. Figure 1-2 shows shares of NNI funding in FY 2006 among the PCAs.

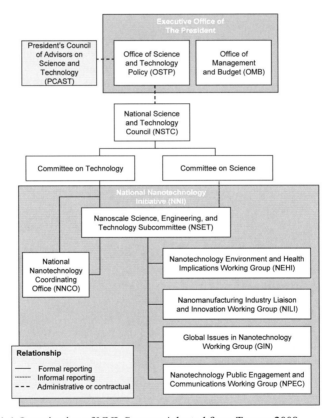

FIGURE 1-1 Organization of NNI. Source: Adapted from Teague 2008.

---

[2]The PCAs are fundamental nanoscale phenomena and processes; nanomaterials; nanoscale devices and systems; instrumentation research, metrology, and standards for nanotechnology; nanomanufacturing; major research facilities and instrumentation acquisition; environmental, health, and safety; and education and ethical, legal, and other societal dimensions (NSET 2007a).

TABLE 1-1 Estimated FY 2008 Agency NNI-Related Investments by Program Component Area (in $ millions)

	Fundamental Nanoscale Phenomena and Processes	Nanomaterials	Nanoscale Devices and Systems	Instrumentation Research, Metrology, and Standards for Nanotechnology	Nanomanufacturing	Major Research Facilities and Instrumentation Acquisition	Environment, Health and Safety	Societal Dimensions	NNI Total[a]
DOD	258.7	68.9	119.8	8.0	5.4	24.6	2.0		487.4
NSF	138.8	62.1	50.3	16.0	26.9	31.6	29.2	33.8	388.7
DOE	51.4	77.5	13.0	12.0	2.0	92.0	3.0	0.5	251.4
DHHS (NIH)	55.6	25.4	125.8	5.9	0.8		7.7	4.6	225.8
DOC (NIST)	22.5	7.4	21.7	16.1	14.4	5.8	0.8		88.7
NASA	1.5	9.7	6.2			0.4	0.2		18.0
EPA	0.2	0.2	0.2				9.6		10.2
DHHS (NIOSH)							6.0		6.0
USDA (FS)	1.3	1.9	1.2	0.4	0.2				5.0
USDA (CSREES)	0.7	1.6	3.1		0.5		0.1	0.1	6.1
DOJ				2.0					2.0
DHS			1.0						1.0
DOT (FHWA)	0.9								0.9
TOTAL	531.6	254.7	342.3	60.4	50.2	154.4	58.5	39.0	1,491.2

[a]Totals may appear to be incorrect because of rounding.
Source: NSET 2008b, Table 3.

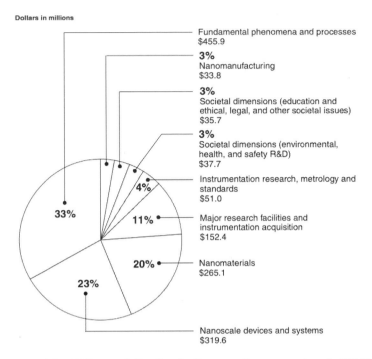

Dollars in millions

Fundamental phenomena and processes
$455.9

3%
Nanomanufacturing
$33.8

3%
Societal dimensions (education and
ethical, legal, and other societal issues)
$35.7

3%
Societal dimensions (environmental,
health, and safety R&D)
$37.7

Instrumentation research, metrology and
standards
$51.0

Major research facilities and
instrumentation acquisition
$152.4

Nanomaterials
$265.1

Nanoscale devices and systems
$319.6

33%
4%
11%
20%
23%

FIGURE 1-2 NNI Research Funding by Program Component Area in FY 2006. Source: GAO 2008.

The PCAs provide a framework that allows the NSET Subcommittee, Office of Science and Technology Policy (OSTP), Office of Management and Budget (OMB), and Congress to be informed of NNI-related activities and that facilitates the management of investments in each PCA and the coordination and direction of nanotechnology-related activities in the participating agencies. The NSET Subcommittee has also established four interagency working groups to address specific cross-agency issues in the context of NNI goals and the PCAs: the Nanotechnology Environment and Health Implications (NEHI) Working Group; the Nanomanufacturing, Industry Liaison, and Innovation Working Group; the Nanotechnology Public Engagement and Communications Working Group; and the Global Issues in Nanotechnology Working Group (see Figure 1-1).

ENVIRONMENT, HEALTH, AND SAFETY

Responsible development of technology can be characterized as the balancing of efforts to maximize the technology's contributions and minimize its adverse consequences. Thus, responsible development of nanotechnology involves an examination of both its applications and its potential implications. It implies a commitment to develop and use technology to meet the most pressing human and societal needs while making every reasonable effort to anticipate and mitigate adverse implications and unintended consequences (NRC 2006).

Nanomaterials have unusual and useful properties. But their unique attributes make them a double-edged sword: although they can be tailored to yield special benefits, they can also have unknown and possibly adverse effects, such as unexpected toxic and environmental effects. The environmental, health, and safety (EHS) implications of nanotechnology are subjects of serious discussion by government agencies and commissions, nongovernment organizations, the research community, industry, insurers, the mass media, and the public. R&D and manufacturing personnel are the ones initially exposed to nanomaterials, so an initial focus of EHS research related to nanomaterials is occupational health and safety risks.

The Growing Importance of Understanding
Environmental, Health, and Safety Issues

The Woodrow Wilson Center Project on Emerging Nanotechnologies reported that 609 consumer products involving nanomaterials were on the market as of April 2008.[3] Some 60% of the consumer products reportedly were in the health and fitness category, which includes skin care and other products designed for direct application to the body (PEN 2008). The consulting firm Lux Research predicts that by 2010 the market value of specific nanomaterials will range from $16 million for nanowires to $1.5 billion for ceramic nanoparticles and that there will be a large expansion in all nanomaterial markets from 2005 to 2010 (Holman 2007). As nanomaterials become incorporated into an increasing number and share of consumer products, opportunities for exposure of workers, the general public, and the environment will also increase, so understanding of the potential risks posed by such exposure takes on greater urgency.

In addition to the application of ceramic and other nanoparticles in cosmetics and skin-care products, expanding applications of nanomaterials with relatively high exposure potential include the use of nanosilver in a wide variety of coatings, clothing, and personal-care products for its antimicrobial properties; use of cerium oxide nanoparticles as catalysts in motor-vehicle fuels; and a variety of ceramic and metallic nanoparticles in coatings (Holman 2007). Applications of carbon nanotubes, ceramic nanoparticles, and metal nanoparticles in composite materials, electronic and optical equipment, and other instruments may offer less exposure potential during the use phase of their life cycle but still result in exposure of workers during manufacturing processes and of workers and the general public at the end of the product life cycle. The combination of the heterogeneity of and enormous variations among nanomaterials and their applications; the

---

[3]These products are identified as nanomaterial-based by the manufacturers or others.

potential for novel forms of toxicity created by their unique size and structural and physical characteristics; and the variations in the frequency, magnitude and duration of releases or exposures, introduces considerable complexity into the design of research programs necessary to understand their potential toxicity.

Researchers, nonprofit organizations, industry, and consumer groups have been calling for an emphasis on EHS research on nanotechnology (Biswas and Wu 2005; Denison 2005; Maynard 2006; Wiesner et al. 2006; Gulledge 2008). In 2004, memorandums from OMB and OSTP to federal-agency heads emphasized the need to give EHS aspects of nanotechnology high priority, noting that "agencies also should support research on the various societal implications of the nascent technology" by placing "a high priority on research on human health and environmental issues...[and] cross-agency approaches" (OMB/OSTP 2004, p.3). The most recent memorandum, for FY 2009, notes that "agencies should strengthen interagency coordination of and support research on potential risks to human health and the environment, consistent with the [NNI (2006)], EHS Research Needs for Engineered Nanoscale Materials" (OMB/OSTP 2007, p. 5).

In 2005, PCAST acknowledged that current knowledge and data for assessing the risks posed by nanotechnology products were incomplete. Furthermore, PCAST said that because exposure to nanomaterials is most likely to occur during manufacturing, research on potential hazards associated with workplace exposure must be given the highest priority (PCAST 2005).[4] In 2005, the NSET Subcommittee formally established the NEHI in a charter that set forth its purpose and objectives (NEHI 2005).[5]

The National Environmental Health Implications Working Group

The NEHI was formed to promote the exchange of information among agencies that support nanotechnology research and those responsible for regulation and guidelines related to nanoproducts; to facilitate identification, priority-setting, and implementation of research needed for the development, use, and oversight of nanotechnology; and to promote communication of information related to research on environmental and health implications of nanotechnology to other government and nongovernment organizations. The NEHI comprises representatives of 18 research and regulatory agencies, OSTP, and OMB and is cochaired by representatives of FDA and the EPA Office of Research and Development.[6]

---

[4]The committee recognizes that PCAST has published a second report, *The National Nanotechnology Initiative: Second Assessment and Recommendations of the National Nanotechnology Advisory Panel* (PCAST 2008a). PCAST assessed the NNI draft strategy in a report, PCAST (2008b), that was an addendum to PCAST (2008a).

[5]The committee recognizes that the informal work of NEHI began as early as 2003.

[6]Current members of NEHI consists of officials from the Consumer Product Safety Commission, Cooperative State Research, Education, and Extension Service, Department of Defense, Department of Energy, Department of State, Department of Transportation,

The NEHI, in its charter, was tasked with the following objectives:

- To improve communication of information related to environmental and health aspects of nanotechnology by the National Nanotechnology Coordination Office (NNCO), the NSET Subcommittee, and individual agencies.
- To assist in the development of information and strategies as a basis for the drafting of guidance in the safe handling and use of nanoproducts by researchers, workers, and consumers.
- To support, with input from the NSET Subcommittee and other appropriate interagency groups, the development of tools and methods for identifying and setting priorities among specific research to enable risk analysis of and regulatory decision-making regarding nanoproducts.
- To support development of nanotechnology standards, including nomenclature and terminology, by consensus-based standards organizations.

The structure of the NEHI mirrors that of the NNI, as it serves primarily as a coordinating body across federal research and regulatory agencies. The NEHI, like the NNI, has no authority over the individual agencies and no budget of its own, so it cannot ensure that agencies address or fund specific kinds of EHS research adequately.

The NEHI's formal coordination efforts have resulted in various reports that have identified EHS research priorities for nanomaterials in a series of documents, starting with *Environmental, Health, and Safety Research Needs for Engineered Nanoscale Materials* (NEHI 2006), which developed five research categories with a total of 75 research needs. The five research categories were instrumentation, metrology, and analytic methods; nanomaterials and human health; nanomaterials and the environment; health and environmental surveillance; and risk-management methods. In 2007, NNI released *Prioritization of Environmental, Health, and Safety Research Needs for Engineered Nanoscale Materials* (NEHI 2007) that reduced the 75 to 25 research needs. In early 2008, *A Strategy for Nanotechnology-Related Environmental, Health, and Safety Research* was released (NEHI 2008).[7] It notes that nanotechnology-related EHS

Environmental Protection Agency, Food and Drug Administration, International Trade Commission, National Aeronautics and Space Administration, National Institute for Occupational Safety and Health, National Institutes of Health, National Institute of Standards and Technology, National Science Foundation, Occupational Safety and Health Administration, Office of Science and Technology Policy, Office of Management and Budget, and U.S. Geological Survey.

[7]In addition to NEHI (2008), many individual agencies have established separate processes to develop their own EHS nanotechnology research strategies. These processes have varied in their structure, their degree of stakeholder involvement, and their complexity. For the most part, the agency personnel engaged in the development of the agency research strategies have been represented on the NEHI, thus allowing for coordination between the NNI research strategy and the individual agency research strategies. The

research and the strategy itself aim to accelerate research to protect public health and the environment and to fill gaps in research, and—in light of the growing level of effort worldwide—to avoid unnecessary duplication of research. The approach, the document notes, is driven by the breadth of issues, from transport in the environment and effects on human health to managing risks and the overarching need to measure and characterize nanomaterials in various environments. Addressing such a variety of issues, the NNI asserts, requires participation by and coordination of the various NNI agencies with their diverse competences and expertise.

STRUCTURE OF THIS REPORT

This National Research Council (NRC) report is an independent assessment of the 2008 NNI document. The NNCO asked the NRC to evaluate the scientific and technical aspects of the draft strategy and to comment in general terms on how the strategy would develop information needed to support the EHS risk-assessment and risk-management needs with respect to nanomaterials.

The committee conducted the evaluation of the NNI draft strategy by asking several questions: What is a research strategy, and more specifically, a risk-research strategy? What are the necessary components of such a strategy (Chapter 2)? Does the strategy have the necessary components (Chapter 3)? For each of the research categories identified in the strategy—including instrumentation, metrology, and analytic methods; human health; environment; exposure assessment; and risk-management methods—are the appropriate research needs identified, is the gap analysis complete and accurate, are priorities among research needs set correctly, and would the research support EHS risk-assessment and risk-management needs (Chapter 4)? Chapter 5 offers the committee's conclusions and recommendations and a look toward future steps in the development of an EHS research strategy for nanomaterials. Society is looking to the scientific community for guidance with respect to nanotechnology, and the committee, in its evaluation, considers nanotechnology to be a field that requires targeted research for understanding the scientific uncertainties surrounding potential EHS risks.

REFERENCES

Biswas, P., and C.Y. Wu. 2005. Nanoparticles and the environment. J. Air Waste Manag. Assoc. 55(6):708-746.
Denison, R.A. 2005. A Proposal to Increase Federal Funding of Nanotechnology Risk Research to at Least $100 Million Annually. Environmental Defense. April 2005 [online]. Available: http://www.edf.org/documents/4442_100milquestionl.pdf [accessed July 29, 2008].

---

individual agency research strategies are of course bounded and shaped by the mission, resources, and regulatory obligations of the agencies developing them.

GAO (General Accountability Office). 2008. Nanotechnology. Better Guidance is Needed to Ensure Accurate Reporting of Federal Research Focused on Environmental, Health, and Safety Risks, March 2008, Washington, DC [online]. Available: http://www.gao.gov/new.items/d08402.pdf.

Gulledge, W. 2008. Presentation at the Second Meeting on Review of the Federal Strategy to Address Environmental, Health, and Safety Research Needs for Engineered Nanoscale Materials, May 5, 2008, Washington, DC.

Holman, M. 2007. Nanomaterials Forecast: Volumes and Applications. Presented at the ICON Nanomaterial Environmental Health and Safety Research Needs Assessment, January 9, 2007, Bethesda, MD [online]. Available: http://cohesion.rice.edu/CentersAndInst/ICON/emplibrary/Nanomaterial%20Volumes%20and%20Applications%20-%20Holman,%20Lux%20Research.pdf [accessed June 19, 2008].

Maynard, A.D. 2006. Nanotechnology: A Research Strategy for Addressing Risk. Project on Emerging Nanotechnology PEN 3. Washington, DC: Woodrow Wilson Center for International Scholars. July 2006 [online]. Available: http://www2.cst.gov.uk/cst/business/files/ww5.pdf [accessed August 22, 2008].

NEHI (Nanotechnology Environmental Health Implications Working Group). 2005. Terms of Reference. Nanotechnology Environmental Health Implications Working Group, Nanoscale Science, Engineering and Technology Subcommittee, Committee on Technology. March 18, 2005.

NEHI (Nanotechnology Environmental Health Implications Working Group). 2006. Environmental, Health, and Safety Research Needs for Engineered Nanoscale Materials. Arlington, VA: National Nanotechnology Coordination Office. September 2006 [online]. Available: http://www.nano.gov/NNI_EHS_research_needs.pdf [accessed Aug. 22, 2008].

NEHI (Nanotechnology Environmental Health Implications Working Group). 2007. Prioritization of Environmental, Health, and Safety Research Needs for Engineered Nanoscale Materials: An Interim Document for Public Comment. Arlington, VA: National Nanotechnology Coordination Office. August 2007 [online]. Available: http://www.nano.gov/Prioritization_EHS_Research_Needs_Engineered_Nanoscale_Materials.pdf [accessed Aug. 22, 2008].

NEHI (Nanotechnology Environmental Health Implications Working Group). 2008. National Nanotechnology Initiative Strategy for Nanotechnology-Related Environmental, Health, and Safety Research. Arlington, VA: National Nanotechnology Coordination Office. February 2008 [online]. Available: http://www.nano.gov/NNI_EHS_Research_Strategy.pdf [accessed Aug. 22, 2008].

NRC (National Research Council). 2002. Small Wonders, Endless Frontiers: A Review of the National Nanotechnology Initiative. Washington, DC: National Academies Press.

NRC (National Research Council). 2006. A Matter of Size: Triennial Review of the National Nanotechnology Initiative. Washington, DC: National Academies Press.

NSET (Nanoscale Science, Engineering, and Technology Subcommittee). 2007a. National Nanotechnology Initiative Strategic Plan. Subcommittee on Nanoscale Science, Engineering, and Technology, Committee on Technology, National Science and Technology Council. December 2007 [online]. Available: http://www.nano.gov/NNI_Strategic_Plan_2007.pdf [accessed June 12, 2008].

NSET (Nanoscale Science, Engineering, and Technology Subcommittee). 2007b. National Nanotechnology Initiative: Research and Development Lending to a Revolution in Technology and Industry: Supplement to the President's FY 2008 Budget.

Arlington, VA: National Nanotechnology Coordination Office. July 2007 [online]. Available: http://www.nano.gov/NNI_08Budget.pdf [accessed Aug. 27, 2008].

NSET (Nanoscale Science, Engineering, and Technology Subcommittee). 2008a. National Nanotechnology Initiative, Subcommittee on Nanoscale Science, Engineering, and Technology, National Science and Technology Council [online]. Available: http://www.nano.gov/index.html [accessed June 9, 2008].

NSET (Nanoscale Science, Engineering, and Technology Subcommittee). 2008b. National Nanotechnology Initiative FY 2009 Budget & Highlights. Subcommittee on Nanoscale Science, Engineering, and Technology, Committee on Technology, National Science and Technology Council [online]. Available: http://www.nano.gov/NNI_FY09_budget_summary.pdf [accessed Aug. 25, 2008].

OMB/OSTP (Office of Management and Budget and Office of Science Technology and Policy). 2004. Updated Administration Research and Development Budget Priorities. M-04-23. Memorandum for the Heads of Executive Departments and Agencies, from John H. Marburger, III, Director, Office of Science and Technology Policy, and Joshua B. Bolten, Director, Office of Management and Budget. August 12, 2004 [online]. Available: http://www.whitehouse.gov/omb/memoranda/fy04/m04-23.pdf [accessed Nov. 19, 2008].

OMB/OSTP (Office of Management and Budget and Office of Science Technology and Policy). 2007. FY 2009 Administration Research and Development Budget Priorities. Memorandum for the Heads of Executive Departments and Agencies, from John H. Marburger, III, Director, Office of Science and Technology Policy, and Stephen S. McMillin, Acting Director, Office of Management and Budget. August 14, 2007 [online]. Available: www.whitehouse.gov/omb/memoranda/fy2007/m07-22.pdf [accessed May 11, 2008].

PCAST (President's Council of Advisors on Science and Technology). 2005. The National Nanotechnology Initiative at Five Years: Assessment and Recommendations of the National Nanotechnology Advisory Panel. May 2005 [online]. Available: http://www.nano.gov/FINAL_PCAST_NANO_REPORT.pdf [accessed Aug. 22, 2008].

PCAST (President's Council of Advisors on Science and Technology). 2008a. The National Nanotechnology Initiative: Second Assessment and Recommendations of the National Nanotechnology Advisory Panel. [online]. Available: http://www.nano.gov/PCAST_NNAP_NNI_Assessment_2008.pdf [accessed November 13, 2008].

PCAST (President's Council of Advisors on Science and Technology). 2008b. Addendum to the National Nanotechnology Initiative: Second Assessment and Recommendations of the National Nanotechnology Advisory Panel. Assessment of the NNI Strategy for Nanotechnology-Related Environmental, Health, and Safety Research. July 2008 [online]. Available: http://www.ostp.gov/galleries/PCAST/PCAST%20Addendum%20Letter.pdf [accessed Jan. 13, 2009].

PEN (Project on Emerging Nanotechnologies). 2008. Consumer Products: An Inventory of Consumer Products Currently on the Market. Project on Emerging Nanotechnologies. [online]. Available: http://www.nanotechproject.org/inventories/consumer/ [accessed June 19, 2008].

Roco, M.C. 2004. The U.S. National Nanotechnology Initiative after 3 years (2001-2003). J. Nanopart. Res. 6(1):1-10.

Roco, M.C., S. Williams, and P. Alivisatos, eds. 2001. Vision for Nanotechnology Research in the Next Decade: Nanotechnology Research Directions. IWGN Workshop Report. Dordrecht: Kluwer.

Siegel, R.W., E. Hu, and M.C. Roco, eds. 1999. Nanostructure Science and Technology: R & D Status and Trends in Nanoparticles, Nanostructured Materials and Nanodevices. Dordrecht: Kluwer.

Teague, E.C. 2008. NNI Strategy for Nanotechnology-Related Environmental, Health, and Safety Research. Presentation at the First Meeting on Review of the Federal Strategy to Address Environmental, Health, and Safety Research Needs for Engineered Nanoscale Materials, March 31, 2008, Washington, DC.

Wiesner, M.R., G.V. Lowry, P. Alvarez, D. Dionysiou, and P. Biswas. 2006. Assessing the risk of manufactured nanomaterials. Environ. Sci. Technol. 40(1):4336-4345.

2

Elements of an Effective Nanotechnology Risk-Research Strategy

OVERVIEW

The term *strategy* is often used to emphasize the importance and relevance of a process, leading to (for example) strategic reports, strategic plans, and strategic research programs. Yet the true meaning of the word has perhaps been lost or diluted through overuse. A possible shift in meaning may be relatively unimportant in many cases. But if there is a need for a well-constructed strategy to address a particular challenge, working from the wrong definition is likely to lead to confusion at best and a poorly conceived plan of action at worst. Therefore, in setting the scene for reviewing the National Nanotechnology Initiative's *Strategy for Nanotechnology-Related Environmental, Health, and Safety Research* (NEHI 2008), it is helpful to think through how the term *strategy* might apply to scientific research in general and to risk-focused research in particular.

Our aim in this chapter is to develop a sense of what the elements of an effective risk-focused research strategy might look like. We start by considering how strategic thinking or planning is related to research in general and what some of the key factors are in developing effective research strategies. We then focus on research aimed specifically at risks to people and the environment—whether real or perceived—and consider aspects of research strategies that are effective in avoiding or reducing the risks. Finally, we propose nine "elements" (see Box 2-1) that we believe are important in developing and implementing an effective research strategy aimed at identifying, assessing, and managing risks associated with nanotechnology. These elements are explained in further detail at the end of the chapter. It is against those elements that *Strategy for Nanotechnology-Related Environmental, Health, and Safety Research* is assessed later.

DEVELOPING EFFECTIVE RESEARCH STRATEGIES

Strategies generally define a set of goals, often in the context of an overarching aim or vision; a plan of action for achieving the goals; and measures for indicating when the goals have been achieved. When that concept is applied to a complex subject, such as scientific research, developing suitable goals, implementable action plans, and measures of success becomes similarly complex. Research is often open-ended and serendipitous, and it can be difficult to formulate goals that will not stifle innovation. Even when the goals are clear—for instance, "to cure cancer" or "to develop renewable energy sources"—the road map for achieving them can be less than obvious. Promising research avenues can lead to dead ends, and seemingly trivial research directions sometimes turn out to be vitally important. Identifying measures of success ahead of time can sometimes seem like staring into a crystal ball. But, as difficult as the process is, strategies are required for science; in which resources are limited and there is a need to justify what is spent on the basis of what is achieved.

Ensuring efficient progress, or "performance," is a key aspect of any research strategy, and selecting useful measures requires a degree of sophistication. In 2002, the Office of Management and Budget designed the Program Assessment Rating Tool (PART) (OMB 2008) in an attempt to evaluate the performance of publicly funded programs, including research and development (R&D) programs. PART does not explicitly address the need for strategies, but it requires agencies to take strategically relevant steps that include defining outcome-based metrics, measuring the efficiency of research programs, and achieving annual efficiency improvements. Applying those steps to scientific research is not easy. The 2008 National Research Council report *Evaluating Research Efficiency in the U.S. Environmental Protection Agency* concluded that "no agency had found a method of evaluating the efficiency of research based on the ultimate outcomes[1] of that research" (p. 10), and indeed the report stated that

---

BOX 2-1 Elements of a Research Strategy

- Vision, or statement of purpose.
- Goals.
- Evaluation of the existing state of science.
- Roadmap.
- Evaluation.
- Review.
- Resources.
- Mechanisms.
- Accountability.

---

[1]Ultimate outcomes include such results as lives saved or clean air and cannot be predicted or known in advance, may occur long after research is completed, and usually depend on action taken by others (NRC 2008).

"ultimate-outcome-based metrics cannot be used to evaluate the efficiency of research" (NRC 2008, p. 5). Rather, the report reflected the need for sophisticated and nuanced approaches to setting and evaluating research agendas in concluding that "the primary goal of research is knowledge, and the development of new knowledge depends on so many conditions that its efficiency must be evaluated in the context of quality, relevance, and effectiveness in addressing current priorities and anticipating future R&D questions" (p. 10). Specifically, the report distinguished between investment efficiency—including the need to identify the most promising lines of research for achieving desired outcomes— and process efficiency, which relates input into research (for example, number of labor hours and dollars spent on laboratory equipment) to what is ultimately achieved.

Development of effective research strategies that generate high-quality, relevant, and effective new knowledge will depend on the nature and context of the work to be done and the decisions to be made. There is a loose hierarchy in how science is organized, from laboratory-level studies through interdisciplinary research programs to governmentwide science initiatives; and different strategies to ensure success are used at each level. Overlying the hierarchy are ideas of how to divide and categorize different "types" of science.

In 1945, Vannevar Bush, director of the Office of Scientific Research and Development, wrote in the report *Science: The Endless Frontier* that "basic research is performed without thought of practical ends. It results in general knowledge and an understanding of nature and its laws. This general knowledge provides the means of answering a large number of important practical problems, though it may not give a complete specific answer to any one of them. The function of applied research is to provide such complete answers" (Bush 1945). The dichotomous perception of basic and applied research has dominated science policy in the United States for much of the last 50 years. Yet as Stokes and others have highlighted, a more nuanced and integrated approach to different "types" of science is perhaps more realistic (Stokes 1997). Rather than use the established but conceptually limited terminology, the panel found it helpful to describe research as "exploratory" or "targeted,"[2] with the understanding that in many cases research will demonstrate attributes associated with both descriptions.

In that context, the overarching aim of exploratory research is the expansion of scientific knowledge, whereas targeted research is focused on achieving specific goals, which are usually practical. The success of exploratory research might be measured with such indicators as an increase in knowledge, and the

---

[2]Similarly, the Environmental Protection Agency uses a nomenclature to describe its research that includes *core research* and *problem-driven research:* problem-driven research is aimed at understanding and solving particular identified environmental problems and reducing associated uncertainties, and core research is aimed at providing broader, more generic information to improve understanding relevant to environmental problems (NRC 1997).

plan of action for a research strategy might include steps to empower the brightest minds to engage in innovative research with as much freedom as possible. In contrast, targeted research has built-in goals, and an implementation plan might consider the best use of multiple mechanisms—contract research, investigator-driven research, or otherwise—to achieve the goals within specific budget and time constraints. In between there is a fruitful crossover regime wherein the ideas underpinning exploratory and targeted research combine, leading to exploratory research that meets real challenges and targeted research that generates knowledge that is not necessarily applied knowledge.

DEVELOPING EFFECTIVE RISK-RESEARCH STRATEGIES

Strategies for risk research—loosely defined as research in support of identifying, assessing, and addressing actual and potential causes of harm to people and the environment—are not typically limited by disciplinary, agency, or philosophic boundaries. They should address challenges of broad societal significance, for example, the reduction or prevention of harm to humans and the environment.

It is the social significance of risk research that perhaps sets it apart from other kinds of research when a research strategy is being developed and implemented. For example, although a poor research strategy for developing new applications might impede progress in a particular field, a poor risk-research strategy has the potential to reverse progress if it results in unanticipated or poorly managed harm to people and the environment. Such a reversal may arise from failure to identify potential risks in a timely manner, failure to understand how to manage new risks effectively, inability to respond to existing risks, or even inability to communicate information on risks effectively. Poor risk-research strategies may also affect perceptions of risk and lead to decision-making in government, business, and society in general that is not necessarily science-based. Ultimately, failure to develop and implement an effective risk-research strategy can potentially lead to economic loss, environmental damage, loss of quality of life, and loss of life itself.

Like any other research strategy, a risk-research strategy will have clearly defined goals, a plan of action for achieving the goals, and measures of success that can inform future modifications of the strategy—all in the context of the existing state of the science. The plan of action for implementing an effective risk-research strategy will rely heavily on targeted research—research that is focused on addressing questions that are critical for ensuring the safety of new materials and products. A long-term risk-research strategy will also encompass exploratory research to generate knowledge that will inform future goals and research directions. With both targeted and exploratory research, useful research will not be limited by conventional disciplines, just as the mechanisms through which materials and products might cause harm do not respect disciplinary boundaries.

Ultimately, the measure of success of a risk-research strategy is the degree to which harm to people and the environment is mitigated or avoided. If the research is in response to an existing problem, success is measured relatively easily as a reduction in the problem (for example, a reduction in lives lost or in the incidence of disease). It is generally acknowledged that risk research ideally is pre-emptive—preventing problems rather than addressing them after the fact—so measures of success are harder to identify. However, it is possible to identify measures that do not rely on prior harm. For instance, in 2006, four research goals to underpin the safety of nanotechnology were identified in a commentary in Maynard et al. (2006)—to develop samplers to detect nanoparticles in air and water, to develop toxicity screening tests, to develop predictive models, and to develop systems for assessing the effects of nanomaterials over their complete life cycle. In each case, it is clear how achieving the goal will help to avoid harmful effects of engineered nanomaterials, and success in achieving the goal is highly measurable.[3] Other approaches to identifying where progress has been made in avoiding harm are possible. And such concepts as "value of information" (which is explored further in Chapter 4) can help to guide limited resources in maximizing the degree to which measurable progress is made (for example, Clemen and Reilly 2004; Yokota and Thompson 2004).

The critical point here is that, hard as it might be to formulate such metrics of success in a risk-research strategy, failure to do so will result in funding of irrelevant research and failure to fund relevant research.

DEVELOPING NANOTECHNOLOGY-SPECIFIC RISK-RESEARCH STRATEGIES

An effective nanotechnology risk-research strategy will be predominantly forward-looking—preparing for potential risks before the technology has a widespread commercial presence. It will address nanotechnology-based products that are beginning to enter commerce and nanotechnologies currently under development. But it will also need to lay the scientific groundwork for addressing future materials and products arising out of new research, new tools, and new cross-fertilization between previously distinct fields of science and technology. The need to be active and forward-looking makes it particularly hard to develop, implement, and evaluate an effective risk-research strategy. In this context, it is helpful to consider briefly how other organizations have approached the challenges of developing such strategies. We aim to highlight some of the approaches taken by others in response to the challenge of developing nanotechnologies safely.

---

[3]For example, the development, commercialization, and adoption by 2010 of instruments that simultaneously measure personal exposure to airborne nanometer-scale particle number, surface area, and mass concentration, as proposed by Maynard et al. (2006), constitute clear goals whose achievement can be quantified against clear time and performance criteria.

In 2004, the British Royal Society and the Royal Academy of Engineering published what has come to be seen as a seminal report on the development of safe and beneficial nanotechnologies. In *Nanoscience and Nanotechnologies: Opportunities and Uncertainties* (Royal Society 2004), the UK carried out a study to define what is meant by *nanoscience* and *nanotechnology*; to summarize and identify gaps in knowledge; to identify potential health and safety, environmental, ethical, and societal effects; and to look toward the future of the field. This study was executed by a working group of experts of diverse backgrounds assembled by the Royal Society and the Royal Academy of Engineering. The report concluded with 21 recommendations to the UK government and other parties on the responsible development of new and emerging technologies. They addressed industrial applications; possible adverse health, safety, and environmental effects; regulatory issues; social and ethical issues; and stakeholder and public discussion. Although the report was not a strategy in itself, it laid the groundwork for developing strategies that would underpin the responsible development of nanotechnologies—including risk research. Three themes in particular stand out among the recommendations: the need for research into what makes nanotechnologies potentially harmful and how to avoid harm throughout their life cycle, the need for research to inform oversight and regulatory decision-making, and the need for independent review of progress in the responsible development of nanotechnologies.

The UK government responded to the report in 2005 with the document *Characterizing the Potential Risks Posed by Engineered Nanoparticles. A First U.K. Government Research Report* (HM Government 2005). It set out a program of research objectives to address potential risks posed by nanoparticles and funding mechanisms to address these objectives with the aim of developing an appropriate framework and measures for controlling unacceptable risks— engineered nanoparticles being the subset of engineered nanomaterials considered to be of most concern (Royal Society 2004). The result was a nanotechnology risk-research strategy that identified what was needed—19 research objectives were identified—and how the UK government proposed to meet the needs.

In 2007, the UK Council for Science and Technology (CST)—the UK government's top-level advisory body on science and technology policy issues—published a 2-year review of progress toward the government's commitments to developing nanotechnology responsibly (CST 2007). The review praised some aspects of the government's progress and criticized others; the details are not as important here as the process. As a result, later in 2007, the government published a second research report, on characterizing the potential risks posed by engineered nanoparticles (HM Government 2007). The second report described progress in addressing the 19 objectives established in 2005, considered where changes in direction and emphasis were needed, addressed issues raised in the CST review, and planned future steps.

It is beyond our scope to evaluate the substance of the UK nanotechnology risk-research plan, but some aspects of the process align with previous discussions on research strategies. The UK government has identified clear aims and objectives, established mechanisms for addressing the objectives, and set in place a process of review and revision. There has been a degree of independence in authoritative input into the research strategy, from the original Royal Society and Royal Academy of Engineering report to the inclusion of nongovernment experts and stakeholders in developing and implementing the strategy.

Looking beyond the UK, the European Union (EU) has been active in identifying and supporting research aimed at addressing potential nanotechnology-related risks. In 2004, the European Commission (EC) released the communication *Towards a European Strategy for Nanotechnology* (EC 2004). The document focused on realizing the societal benefits of nanotechnology, but it emphasized addressing potential risks in informed decision-making: "Nanotechnology must be developed in a safe and responsible manner. Ethical principles must be adhered to and potential health, safety or environmental risks scientifically studied, . . . in order to prepare for possible regulation" (p. 3).

After the 2004 communication, the EC published *Nanosciences and Nanotechnologies: An Action Plan for Europe 2005–2009* (EC 2005) as a communication to the Competitiveness Council, the European Parliament, and the Economic and Social Committee. In the action plan, the EC recommended EU and member-state actions to address eight elements of nanotechnology development, including public health, safety, and environmental and consumer protection (action point 6). Key to that action point were commitments and recommendations to identify and address safety concerns, evaluate and minimize exposures, and ensure adequate oversight of nanotechnologies—in essence, to establish a framework for strategic research that led to informed decisions. Like the UK nanotechnology plan, the EC plan provided for regular review, and in 2007 the EC published its first implementation report on the action plan (EC 2007).

Although the European action plan for nanotechnology did not explicitly include a risk-research strategy, it did provide a framework for developing such strategies. In testimony to the committee from representatives of the EU directorate general for science, research, and development and the directorate general for health and consumer affairs (Aguar 2008; Martin 2008), it was clear that the EU response to developing nanotechnologies responsibly involves a complex interplay between EU agencies, member states, and nongovernment stakeholders. There does not appear to be a single overarching strategy governing risk research in Europe, but rather multiple initiatives that together form a cohesive approach to supporting research that will inform policy decisions. Two initiatives in particular highlight the current state of affairs: the European Union Seventh Framework Program for Research and Development and a review of risk-assessment methods for assessing the risks associated with nanomaterials conducted by the Scientific Committee on Emerging and Newly Identified Health Risks (SCENIHR). The SCENIHR is an independent scientific committee established to provide the EC with sound scientific advice for preparing policy and proposals related to public health and the environment. It is one of three such committees that address nonfood issues; it complements the Scientific Committee on Consumer Products and the Scientific Committee on Health and Environmental Risks.

Research at the EU level is funded through framework programs that establish the aims and aspirations of pan-European R&D initiatives. The current program is Framework Program 7 (FP7) and will run from 2007 to 2013 (EC 2006). Over this period, over €3.5 billion will be invested in nanotechnology R&D, some of which will be invested in risk research. Calls for proposals within the framework program range from enabling exploratory research to targeting specific issues and typically require collaboration between disciplines, countries, and public and private organizations. In the 2007 call for proposals, four categories focused specifically on environmental health and safety: portable devices for exposure measurement and analysis, risk assessment of engineered nanoparticles, review of the scientific literature on potential risks, and creation of a critical database on the effects of nanoparticles on the environment, health, and safety. Those topic categories, although forming only a small part of the research needed to address potential adverse effects of nanotechnologies, targeted specific issues identified through consultation with a broad base of experts and stakeholders. This process of consultation is continuing to inform research calls under FP7.

The second initiative of interest here is an "opinion" published by the SCENIHR in 2007 (SCENIHR 2007). The SCENIHR was asked, in light of current scientific knowledge and in relation to the general information on and practices of chemical risk assessment, to assess the appropriateness of risk-assessment methods described in the current chemical-related technical guidance documents for risk assessment of nanomaterials and to suggest improvements in the method. Although it did not result in a risk-research strategy, the assessment was important on three counts: it formed part of the tapestry of independent and expert science-based input into the EU planning and decision-making process, which includes strategic decision-making on research directions; it systematically established the level of information needed on emerging nanomaterials to evaluate—and thus manage—potential risks and in doing so provided a framework for developing research strategies to fill gaps; and it explicitly identified research subjects that need further attention if informed decisions were to be made on responsible development and use of nanomaterials.

Those two examples and others not included here are indicative of an approach to risk research in Europe that engages a broad array of experts and stakeholders, identifies key policy goals, establishes mechanisms for supporting research to address the goals, and periodically reviews progress toward the goals.

The Organisation for Economic Co-operation and Development (OECD) has also begun to address the coordination of nanotechnology risk-research strategies among member countries. In 2006, the Working Party on Manufactured Nanomaterials (WPMN) was established under the OECD Chemicals Committee with the aim of promoting international cooperation in aspects of manufactured nanomaterials related to human health and environmental safety to assist in the development of rigorous safety evaluation of nanomaterials. The working party is supporting eight projects that collate, coordinate, and disseminate information and activities linking scientific understanding to the effective oversight of engineered nanomaterials; the second project addresses research strategies regarding manufactured nanomaterials.

Although the OECD WPMN is not developing a nanotechnology risk-research strategy, its aim is to exchange information and identify common research needs to address human-health and environmental-safety issues associated with manufactured nanomaterials (or engineered nanomaterials) and to undertake to meet the needs. In many ways, that is a step toward establishing an international framework within which individual countries and economies can develop risk-research strategies that address the needs of decision-makers while being coordinated with other global initiatives. The OECD process predominantly involves government representatives, but there are provisions in the organization's structure for industry and nongovernment environmental organizations to participate in the working party. It is thus likely that when the results of the research-strategies project begin to emerge, they will to some extent represent input from stakeholders beyond government departments and agencies. However, it should be recognized that non-government stakeholder involvement in this process is neither inclusive nor representative.

Apart from national and international government initiatives to develop nanotechnology risk-research strategies, there have been a number of independent initiatives to map out strategic research needs and approaches. Several papers have been published in recent years highlighting specific research needs, including *Principles for Characterizing the Potential Human Health Effects from Exposure to Nanomaterials: Elements of a Screening Strategy* (Oberdörster et al. 2005), *Safe Handling of Nanotechnology* (Maynard et al. 2006), and *Hazard Assessment for Nanoparticles—Report from an Interdisciplinary Workshop* (Balbus et al. 2007).

Recently, the International Council on Nanotechnology released *Towards Predicting Nano-Bio Interactions: An International Assessment of Research Needs for Nanotechnology Environment, Health and Safety* (ICON 2008). It reports on two international multistakeholder workshops that were tasked to identify and set priorities for the research needed to classify nanomaterials by physical and chemical properties and to develop predictive models for their interactions with living systems. The result was 36 recommendations on research needed to understand more fully how nanomaterials interact with biologic systems and on how to use this knowledge to avoid undue harm on near-term, middle-term and long-term time scales.

A comprehensive overview of challenges to and solutions for developing a nanotechnology risk-research strategy was published by the Project on Emerging Nanotechnologies (Maynard 2006). *Nanotechnology: A Research Strategy for Addressing Risk* draws on nine published reports,[4] including the Royal Society and Royal Academy of Engineering report (Royal Society 2004) and the EC action plan for nanoscience and nanotechnologies published in 2005 (EC 2005), and develops recommendations on the aims, objectives, and implementation of a responsive risk-research strategy. The report differs from others cited here in that it is one person's opinion rather than reflecting the views of multiple stakeholders and experts. However, it draws heavily on opinions and perspectives published elsewhere.

Maynard (2006) identifies "the roll-out of 'safe' nanotechnologies" as the overarching aim of a risk-research strategy and identifies a number of research objectives, including addressing human and environmental health hazards, material characterization and exposure, exposure control, and risk reduction. It considers how the objectives might be best achieved in a timely manner by developing and implementing an effective research strategy. In particular, four components of a government-led strategic research framework are identified and expanded on: linking research to oversight, balancing different approaches to research and research funding (specifically, balancing exploratory and targeted research and using the full spectrum of funding mechanisms appropriately), ensuring authority to direct research, and enabling coordination and partnerships.

Much of the report stresses the importance of targeted research in an effective strategy, which would lead to informed decision-making, but it also stresses the need for exploratory research that will underpin future targeted questions regarding emerging risks. In addition, the report distinguishes between research that addresses nanotechnology risks directly and what it refers to as "indirect research." The latter is identified as research that has the potential to inform an understanding of the effects of nanotechnologies but is not necessarily directed primarily at risks. For example, research into general characterization methods or research into nanotechnology-based drug development might be considered indirect research in the context of risk but lead to risk-relevant information. The report attaches considerable importance to that category of research but warns that "unless this latent potential [is] realized through targeted research, the work will be worthless to understanding and addressing risk."

On the basis of those examples and others not included in this brief overview, it is fair to say that an understanding of what an effective nanotechnology risk-research strategy might look like is still evolving. However, common themes emerge from the above examples and discussions, including the need to link research to decision-making processes, to identify overarching aims and key objectives, to ensure broad expert and multistakeholder input, to ensure access to adequate resources, and to initiate a program of independent review and revision.

---

[4]Royal Society (2004); Chemical Industry Vision 2020 Technology Partnership and SRC (2005); Dennison (2005); EC (2005); EPA (2005); HM Government (2005); Maynard and Kuempel (2005); NIOSH (2005); and Oberdörster et al. (2005).

ELEMENTS OF A RISK-RESEARCH STRATEGY

On the basis of the preceding discussion of research strategies in general and nanotechnology risk-research strategies in particular, the present committee suggests nine elements as key components of an effective research strategy that addresses environmental, health, and safety effects of emerging nanotechnologies. The importance of those elements will depend on the context of a given research strategy. However, it is hard to imagine a successful risk-research strategy that does not address each one of them to some extent. Consequently, the elements have informed our assessment of the National Nanotechnology Initiative *Strategy for Nanotechnology-Related Environmental, Health, and Safety Research* (NEHI 2008).

The nine elements are the following:

- *Vision, or statement of purpose.* What is the ultimate purpose of conducting research on potential risks associated with nanotechnology?
- *Goals.* What specific research goals need to be achieved to guide the development and implementation of nanotechnologies that are as safe as possible?
- *Evaluation of the state of science.* What is known about the potential for the products of nanotechnology to cause harm and about how possible risks might be managed? Could existing knowledge and expertise be mined to provide insight into and solutions to potential nanotechnology-related risks?
- *Road map.* What is the plan of action to achieve the stated research goals? What are the specific objectives, and when do they need to be achieved? How will available resources, institutions, and funding mechanisms be used? Are there needs for new mechanisms to ensure that the right research is carried out? How will other efforts and initiatives be leveraged, including industry and international initiatives? How will the road map be adjusted in light of new knowledge? What is the time required for the plan to become effective?
- *Evaluation.* How will research progress be measured, and who will be responsible for measuring it? Are there measurable milestones that can be evaluated against a clear timeline?
- *Review.* How will the strategy be revised in light of new findings, to ensure that it remains responsive to the overarching vision and goals?
- *Resources.* Are there sufficient resources to achieve the stated goals? If not, what are the plans to obtain new resources or to leverage other initiatives to achieve the goals?
- *Mechanisms.* What are the most effective approaches to achieving the stated goals? How will exploratory and targeted research be used? What will be the balance between principal-investigator–driven and goal-driven research and between intramural and extramural research programs? How will research efforts be coordinated to ensure a coherent approach to achieving stated goals?

What provisions are there for enabling interdisciplinary research that crosses established funding and agency boundaries?

- *Accountability.* How will stakeholders participate in the process of developing and evaluating a research strategy? Who will be accountable for progress toward stated goals? Who will be responsible for disseminating information generated within the research strategy and ensuring its use in raising awareness and making decisions?

REFERENCES

Aguar, P. 2008. The EU Framework for EHS Research on Nanotechnology. Presentation at the Second Meeting on Review of the Federal Strategy to Address Environmental, Health, and Safety Research Needs for Engineered Nanoscale Materials, May 5, 2008, Washington, DC.

Balbus, J.M., A.D. Maynard, V.L. Colvin, V. Castranova, G.P. Daston, R.A. Denison, K.L. Dreher, P.L. Goering, A.M. Goldberg, K.M. Kulinowski, N.A. Monteiro-Riviere, G. Oberdörster, G.S. Omenn, K.E. Pinkerton, K.S. Ramos, K.M. Rest, J.B. Sass, E.K. Silbergeld, and B.A. Wong. 2007. Hazard assessment for nanoparticles: Report from an Interdisciplinary Workshop. Environ. Health Perspect. 115(11):1654-1659.

Bush, V. 1945. Science - The Endless Frontier. Washington, DC: Office of Scientific Research and Development [online]. Available: http://www.nsf.gov/about/history/vbush1945.htm [accessed July 3, 2008].

Chemical Industry Vision 2020 Technology Partnership and SRC (Semiconductor Research Corporation). 2005. Joint NNI-ChI CBAN and SRC CWG5 Nanotechnology Research Needs Recommendations [online]. Available: http://www.chemicalvision2020.org/pdfs/chcm-scmi_ESH_recommendations.pdf [accessed Aug. 26, 2008].

Clemen, R.T., and T. Reilly. 2004. Making Hard Decisions with Decision Tools, 2nd Ed. Florence, KY: Brooks/Cole Publishers.

CST (Council for Science and Technology). 2007. Nanosciences and Nanotechnologies: A Review of Government's Progress on its Policy Commitments. London, UK: Council for Science and Technology [online]. Available: http://www.oecd.org/dataoecd/58/60/38390159.pdf [accessed Aug. 26, 2008].

Denison, R.A. 2005. A Proposal to Increase Federal Funding of Nanotechnology Risk Research to at Least $100 Million Annually. Environmental Defense. April 2005 [online]. Available: http://www.edf.org/documents/4442_100milquestionl.pdf [accessed July 29, 2008].

EC (European Commission). 2004. Communication from the Commission: Towards a European Strategy for Nanotechnology. COM(2004) 338 final. Brussels: Commission of the European Communities [online]. Available: ftp://ftp.cordis.europa.eu/pub/nanotechnology/docs/nano_com_en.pdf [accessed Aug. 26, 2008].

EC (European Commission). 2005. Communication from the Commission to the Council, the European Parliament and the Economic and Social Committee: Nanoscience and Nanotechnologies: An Action Plan for Europe 2005 - 2009. COM (2005) 243 final. Brussels: Commission of the European Communities [online]. Available: ftp://ftp.cordis.europa.eu/pub/nanotechnology/docs/nano_action_plan2005_en.pdf [accessed Aug. 26, 2008].

EC (European Commission). 2006. FP7 – Tomorrow's Answers Start Today. Seventh
    Framework Programme for Research and Technological Development, European
    Communities [online]. Available: http://ec.europa.eu/research/fp7/pdf/fp7-factshee
    ts_en.pdf [accessed Oct. 9, 2008].
EC (European Commission). 2007. Communication from the Commission to the Council,
    the European Parliament and the European Economic and Social Committee:
    Nanosciences and Nanotechnologies: An Action Plan for Europe 2005-2009. First
    Implementation Report 2005-2007. COM (2007) 505 final. Brussels: Commission
    of the European Communities [online]. Available: ftp://ftp.cordis.europa.eu/pub/
    nanotechnology/docs/com_2007_0505_f_en.pdf [accessed January 12, 2009].
EPA (U.S. Environmental Protection Agency). 2005. Nanotechnology White Paper. Ex-
    ternal Review Draft. Science Policy Council, U.S. Environmental Protection
    Agency, Washington, DC. December 2, 2005 [online]. Available: http://www.epa.
    gov/OSA/pdfs/EPA_nanotechnology_white_paper_external_review_draft_12-02-
    2005.pdf [accessed Aug. 26, 2008].
HM Government. 2005. Characterizing the Potential Risks Posed by Engineered
    Nanoparticles: A First UK Government Research Report. Department for Envi-
    ronment, Food and Rural Affairs, London [online]. Available: http://www.
    defra.gov.uk/Environment/nanotech/research/pdf/nanoparticles-riskreport.pdf [ac-
    cessed Aug. 26, 2008].
HM Government. 2007. Characterizing the Potential Risks Posed by Engineered Nano-
    particles: A Second UK Government Research Report. Department for Environ-
    ment Food and Rural Affairs, London [online]. Available: http://www.defra.
    gov.uk/environment/nanotech/research/pdf/nanoparticles-riskreport07.pdf [access-
    ed Aug. 26, 2008].
ICON (International Council on Nanotechnology). 2008. Towards Predicting Nano-
    Biointeractions: An International Assessment of Nanotechnology Environment,
    Health, and Safety Research Needs. International Council on Nanotechnology No.
    4. May 1, 2008 [online]. Available: http://cohesion.rice.edu/CentersAndInst/IC
    ON/emplibrary/ICON_RNA_Report_Full2.pdf [accessed Aug. 26, 2008].
Martin, P. 2008. On the Importance of Nanotechnologies Safety Research. Presentation at
    Second Meeting on Review of the Federal Strategy to Address Environmental,
    Health, and Safety Research Needs for Engineered Nanoscale Materials, May 5,
    2008, Washington, DC.
Maynard, A.D. 2006. Nanotechnology: A Research Strategy for Addressing Risk. Project
    on Emerging Nanotechnology PEN 3. Washington, DC: Woodrow Wilson Center
    for International Scholars. July 2006 [online]. Available: http://www2.cst.gov.uk/
    cst/business/files/ww5.pdf [accessed August 22, 2008].
Maynard, A.D., and E.D. Kuempel. 2005. Airborne nanostructured particles and occupa-
    tional health. J. Nanopart. Res. 7(6):587-614.
Maynard, A.D., R.J. Aitken, T. Butz, V. Colvin, K. Donaldson, G. Oberdörster, M.A.
    Philbert, J. Ryan, A. Seaton, V. Stone, S.S. Tinkle, L. Tran, N.J. Walker, and D.B.
    Warheit. 2006. Safe handling of nanotechnology. Nature 444(7117):267-269.
NEHI (Nanotechnology Environmental Health Implications Working Group). 2008. Na-
    tional Nanotechnology Initiative Strategy for Nanotechnology-Related Environ-
    mental, Health, and Safety Research. Arlington, VA: National Nanotechnology
    Coordination Office. February 2008 [online]. Available: http://www.nano.gov/NNI
    _EHS_Research_Strategy.pdf [accessed Aug. 22, 2008].
NIOSH (National Institute for Occupational Safety and Health). 2005. Strategic Plan for
    NIOSH Nanotechnology Research: Filling the Knowledge Gaps. Draft, September

28, 2005. National Institute for Occupational Safety and Health [online]. Available: http://www.cdc.gov/niosh/topics/nanotech/strat_plan.html [Aug. 27, 2008].

NRC (National Research Council). 1997. Building a Foundation for Sound Environmental Decisions. Washington, DC: National Academy Press.

NRC (National Research Council). 2008. Evaluating the Efficiency of Research and Development Programs at EPA. Washington, DC: The National Academies Press.

Oberdörster, G., A. Maynard, K. Donaldson, V. Castranova, J. Fitzpatrick, K. Ausman, J. Carter, B. Karn, W. Kreyling, D. Lai, S. Olin, N. Monteiro-Riviere, D. Warheit, and H. Yang; ILSI Research Foundation/Risk Science Institute Nanomaterial Toxicity Screening Working Group. 2005. Principles for characterizing the potential human health effects from exposure to nanomaterials: Elements of a screening strategy. Part Fibre Toxicol. 2(1):8 doi:10.1186/1743-8977-2-8 [online]. Available: http://www.particleandfibretoxicology.com/content/2/1/8 [accessed Aug. 27, 2008].

OMB (Office of Management and Budget). 2008. Program Assessment Rating Tool (PART). Office of Management and Budget [online]. Available: http://www.johnmercer.com/omb part.htm [accessed Aug. 27, 2008].

Royal Society. 2004. Nanoscience and Nanotechnologies: Opportunities and Uncertainties. London: The Royal Society & the Royal Academy of Engineering [online]. Available: http://www.nanotec.org.uk/finalReport.htm [accessed Aug. 27, 2008].

SCENIHR (Scientific Committee on Emerging and Newly-Identified Health Risks). 2007. Opinion On the Appropriateness of the Risk Assessment Methodology in Accordance with the Technical Guidance Documents for New and Existing Substances for Assessing the Risks of Nanomaterials. Brussels: European Commission [online]. Available: http://ec.europa.eu/health/ph_risk/committees/04_scenihr/docs /scenihr_o_010.pdf [accessed Aug. 27, 2008].

Stokes, D.E. 1997. Pasteur's Quadrant: Basic Science and Technological Innovation. Washington, DC: Brookings Institution.

Yokota, F., and K.M. Thompson. 2004. Value of information analysis in environmental health risk management decisions: Past, present, and future. Risk Anal. 24(3):635-650.

3

Evaluation of the Federal Strategy

In Chapter 2, the committee identified the key elements of a nano-risk research strategy: an evaluation of the existing state of science, an overarching vision or statement of purpose, goals to ensure safe development of nanotechnologies, a road map for ensuring achievement of stated goals, evaluation for assessing progress in achieving the goals, a process of review to ensure the strategy remains responsive to the overarching vision and goals, identification of resources, mechanisms to achieve goals, and accountability. The committee evaluated *Strategy for Nanotechnology-Related Environmental, Health, and Safety Research* (NEHI 2008) by considering whether it contained those elements. In its evaluation, the committee considered input from public sessions held at the National Academies (March 31 and May 5, 2008) at which representatives of the Nanotechnology Environmental and Health Implications Working Group (NEHI) and of the stakeholder community—including industry, nongovernment organizations, and the insurance sector—provided comments on the federal strategy. Many of the stakeholders' comments echoed sentiments of the committee and are provided here as support for the committee's views on NNI (NEHI 2008). (See Appendix C for an agenda of the public sessions.)

The committee concluded that the development of the NNI (NEHI 2008) has provided a unique opportunity for coordination, planning, and consensus-building among 18 agencies within NEHI. However, the committee determined that the NNI document does not have the essential elements of a nano-risk research strategy, inasmuch as it does not evaluate the state of science, does not contain a clear set of goals, and does not have a plan of action for achieving the goals or mechanisms to review and evaluate funded research and assess whether progress has been achieved. There is no attempt to show how existing research will lead to answers to critical questions that the federal government, the research community, and other stakeholders are grappling with.

IS THERE AN EVALUATION OF THE EXISTING STATE OF SCIENCE?

The research categories and needs presented in the strategy are based on priorities reviewed and evaluated in the previous NNI reports, *Environmental, Health, and Safety Research Needs for Engineered Nanoscale Materials* (NEHI 2006) and *Prioritization of Environmental, Health, and Safety Research Needs for Engineered Nanoscale Materials: An Interim Document for Public Comment* (NEHI 2007), both of which received public comment. The first of those reports developed five research categories with a total of 75 research priorities. The priorities were reduced to 25 in the second report. The new strategy (NEHI 2008) attempts to develop timelines and sequence the research needs and uses an accounting of research projects of FY 2006 to determine the strengths, limitations, and data gaps of the research portfolio.

There is no evaluation of the existing state of science or of federally funded research in each of the five categories identified in the strategy—instrumentation, metrology, and analytic methods; nanomaterials and human health; nanomaterials and the environment; human and environmental exposure assessment; and risk-management methods. Rather, the research categories and identified research needs (see Box 3-1) are analyzed solely in the context of FY 2006 research projects. The committee questions the NNI's use of FY 2006 data to assess the extent to which federally funded environmental, health, and safety (EHS) research for nanomaterials is supporting the selected research needs. The majority of the research projects listed for FY 2006 focused on fundamentals of nanoscience that are not explicitly associated with risk, or on developing nano-technology applications.[1] There also is no clear connection between the research projects and how they will inform an understanding of risk. Without a clear articulation of how the research projects will inform that understanding, the report's assessment is highly misleading and inappropriately used to identify whether research needs are being addressed.

NNI (NEHI 2008) contains conflicting statements about the use of FY 2006 research projects to evaluate research needs. The document states that "this analysis of strengths, weaknesses, and gaps will inform agency decisions about the magnitude and balance of future EHS research investments" (NEHI 2008, p. 9). But the document continues, "data gathered for FY 2006 represent a one-time-only 'snapshot' of the NNI agencies' EHS research portfolios in one year. However, these are likely to be indicative of the overall trends in agency investments in more recent years" (NEHI 2008, p. 9). The strategy goes on to acknowledge limits of the gap analysis, including statements that the data represent only projects funded in FY 2006; that the data represent planned research, not research results; and that only federally funded research is accounted for—there is no mention of research funded by industry, nonprofit organizations, or other countries. Those statements in the strategy were echoed by Altaf Carim,

---

[1]The 246 FY 2006 research projects listed include research on instrumentation and metrology and on medical applications that is not captured in the list of 130 environmental, health, and safety research projects included in the annual supplement to the president's budget (Teague, unpublished material, 2008).

program manager in the Office of Science, Department of Energy, who acknowledged in his written testimony to the committee "that data was one of the *inputs* to the planning process—a snapshot of Federal activity that in fact was analyzed in order to determine where there were gaps and to identify the priority areas for future investment" (Carim 2008, p.1).

BOX 3-1 Priority Environmental, Health, and Safety Research Needs for Engineered Nanoscale Materials, as Identified in the 2008 National Nanotechnology Initiative Research Strategy

Instrumentation, Metrology, and Analytical Methods

1. Develop methods to detect nanomaterials in biological matrices, environment, and workplace.
2. Understand how chemical and physical modifications affect the properties of nanomaterials.
3. Develop methods for standardizing assessment of particle size, size distribution, shape, structure, and surface area.
4. Develop certified reference materials for chemical and physical characterization of nanomaterials.
5. Develop methods to characterize a nanomaterial's spatio-chemical composition, purity, and heterogeneity.

Nanomaterials and Human Health

Overarching Research Priority: Understand generalizable characteristics of nanomaterials in relation to toxicity in biological systems.

Broad Research Needs:

- Understand the absorption and transport of nanomaterials throughout the human body.
- Develop methods to quantify and characterize exposure to nanomaterials and characterize nanomaterials in biological matrices.
- Identify or develop appropriate *in vitro* and *in vivo* assays/models to predict *in vivo* human responses to nanomaterials exposure.
- Understand the relationship between the properties of nanomaterials and uptake via the respiratory or digestive tracts or through the eyes or skin, and assess body burden.
- Determine the mechanisms of interaction between nanomaterials and the body at the molecular, cellular, and tissular levels.

Nanomaterials and the Environment

1. Understand the effects of engineered nanomaterials in individuals of a species and the applicability of testing schemes to measure effects.

(Continued)

BOX 3-1 Continued

2. Understand environmental exposures through identification of principle sources of exposure and exposure routes.
3. Evaluate abiotic and ecosystem-wide effects.
4. Determine factors affecting the environmental transport of nanomaterials.
5. Understand the transformation of nanomaterials under different environmental conditions.

Human and Environmental Exposure Assessment

1. Characterize exposures among workers.
2. Identify population groups and environments exposed to engineered nanoscale materials.
3. Characterize exposure to the general population from industrial processes and industrial and consumer products containing nanomaterials.
4. Characterize health of exposed populations and environments.
5. Understand workplace processes and factors that determine exposure to nanomaterials.

Risk Management Methods

Overarching Research Priority: Evaluate risk management approaches for identifying and addressing risks from nanomaterials.

1. Understand and develop best workplace practices, processes, and environmental exposure controls.
2. Examine product or material life cycle to inform risk reduction decisions.
3. Develop risk characterization information to determine and classify nanomaterials based on physical or chemical properties.
4. Develop nanomaterial-use and safety-incident trend information to help focus risk management efforts.
5. Develop specific two-way risk communication approaches and materials.

Source: NEHI 2008.

The committee's concerns about the limitations of the assessment of the state of science were reflected by Carolyn Cairns, program leader of product safety for Consumer's Union, at the May 5, 2008 workshop: "The document resembles a laundry list of ad hoc projects that some agencies have shoe-horned into relevance for environmental health and safety. It is not a strategy that will accelerate the research needed to prevent our toxic past from repeating itself in nano-form. The document fails to articulate how the disparate projects outlined will be pulled together to glean meaningful conclusions that participating agencies can use to protect the public from dangers inherent in commercializing nanomaterials" (Cairns 2008, p.1).

DOES THE STRATEGY HAVE A VISION OR STATED PURPOSE?

The strategy document has various statements of purpose, but none provides a clear vision of where understanding of the environmental, health, and safety implications of nanotechnology should be in 5 or 10 years, including ensuring that the results of research are useful and applicable to decision-making for reducing potential environmental, health, and safety effects of nanomaterials. Relevant research is also needed for policy decisions on government oversight, in industry, and in a broader societal context.

The statement that stands out most as the purpose of the strategy document is that "the NEHI Working Group developed this nanotechnology-related EHS research strategy to accelerate progress in research to protect public health and the environment, and to fill gaps in, and—with the growing level of effort worldwide—to avoid unnecessary duplication of, such research" (NEHI 2008, p. 1). That statement is adequate for an open-ended research program with no definite objectives, but it stops short of ensuring that the results of strategic research are useful and applicable to decision-making that will reduce the potential environmental and health effects of nanotechnologies.

The committee notes that in some cases the strategy document reads as though it has two stated objectives: continuing to support nanotechnology and understanding risks. As the strategy states, "this effort has entailed identifying and prioritizing EHS research for nanomaterials; analyzing the current research portfolio in detail; performing a gap analysis to determine areas requiring emphasis; and developing a strategy to address these areas and *to sustain the diverse program aimed at advancing knowledge and supporting risk decision making*" (NEHI 2008, p.1; emphasis added). Those two objectives are emphasized again: "the NNI aims to maximize the benefits of this new technology at the same time it is developing an understanding of any potential risks and means to manage such risks" (NEHI 2008, p. 1). Stakeholders at the committee's May 5, 2008 public session expressed concerns, similar to those of the committee, that the strategy document seemed to be divided between protecting public health from potential risks of nanomaterials and developing nanotechnology products.

A clear and distinct vision may be difficult for the coordinating agencies to articulate and agree to inasmuch as they reflect different backgrounds, goals, and legislative mandates (see discussion on limitations of the NNI and the NEHI at the end of this chapter).

DOES THE STRATEGY HAVE GOALS TO ENSURE THE SAFE DEVELOPMENT OF NANOTECHNOLOGIES, AND IS THERE A ROAD MAP FOR ACHIEVING STATED GOALS?

NNI (NEHI 2008) does not present goals or a plan of action for achieving them. Although it identifies five "research needs" for each of the five general categories (see Box 3-1), the needs are not articulated as clear goals. There also are no measures of progress to evaluate how and to what extent the goals are being attained. As William Kojola, AFL-CIO industrial hygienist, commented, "a comprehensive set of goals and objectives should first be identified and then a strategy needs to be developed to accomplish these goals and objectives. . . . The current NNI strategy appears to essentially consist of a listing of agency projects cobbled together to look like a strategy" (Kojola 2008, p. 2).

The committee recognizes that the "emphasis diagrams" (NEHI 2008, Figures 3, 5, 7, 9, and 11) for the research needs in the five categories provide some element of timeframe and sequencing. As the strategy states, "priority. . . was considered both in terms of the kind of information developed (some information is of greater relevance than others to supporting risk management) and the appropriate sequencing of research (some research should be timed to occur following other research in order to gain the greatest benefit to decision making with respect to product use, regulations, and conduct of research)" (NEHI 2008, p. 10). Some research needs (for example, in the category of instrumentation, metrology, and analytic methods) could be translated into measureable objectives, but for many others there are insufficient details to determine the measurable objectives.

A key element of any strategy is to identify goals and measures of progress or success *before* assessing what is being done. That allows a clear assessment of the value of current activities, whether in the organization—the government in this case—or outside it (such as research supported by industry, nonprofits, or other countries). Such an approach enables development of an action plan to leverage other efforts and to address and measure research deficiencies in a way that is transparent.

Because NNI (NEHI 2008) does not establish goals and a plan of action, there is no roadmap; the document never raises such questions as, What other research activities should be leveraged? and What additional research activities are needed? Rather, it asserts that current activities are addressing research needs. Terry Medley, global director of corporate regulatory affairs at DuPont, highlighted in his May 5, 2008 presentation to the committee the need for metrics for evaluation as a critical component of successful implementation of the NNI strategy (Medley 2008).

The committee notes that the role of goals and milestones in a complex and emerging research field is not to predict and hold research organizations to the predictions but to map out a systematic plan with chartable actions, which will of necessity change. The committee recognizes that useful goals and a plan of action in this context are not easy to formulate, but they are urgently needed.

DOES THE STRATEGY PROVIDE FOR EVALUATION OF RESEARCH PRIORITIES AND AN ASSESSMENT OF RESEARCH PROGRESS?

NNI (NEHI 2008) states that "the task forces analyzed the portfolio of projects in each category to determine the balance of effort. . . . In addition to tabulating the number of projects and total funding . . . the task forces considered the breadth of research, such as variety of nanomaterials or routes of exposure" (NEHI 2008, p. 7). Although there is some justification in the document for the research priorities selected, it is marginal. The research priorities were developed in NNI (NEHI 2006) and NNI (NEHI 2007), but it is not clear from those documents how they were ultimately selected.[2] It is also not possible to discern relative priorities among the various research needs shown in each of the five categories or even among the five categories. Although the strategy clearly states that no effort was made to set priorities among the categories, because the category of instrumentation, metrology, and analytic methods is cross-cutting—supporting research in every other category—it has high priority itself (NEHI 2008, p. 9).

In general, the process behind the selection of the research priorities and the later priority weightings in the emphasis diagrams is not transparent. There also is little discussion of the itemized research needs in the emphasis diagrams. Many of the research needs make sense, but a few are questionable. For instance, why put the development of materials to support exposure assessment before the development of materials to support toxicology studies (NEHI 2008, p. 18)? Why delay research into alternative surface-area measurement methods for 10 years in light of its being identified as a critical research subject (NEHI 2008, p. 18)? Why delay the development of high-throughput screening methods by 5 years (NEHI 2008, p. 24)? There are many other examples. Further discussion of research priority-setting in each of the research categories is discussed in Chapter 4.

Without clear goals, as discussed above, effective priority-setting is nearly impossible. Without effective priority-setting among research needs, measurement of research progress makes little sense.

---

[2]NNI (NEHI 2006) and NNI (NEHI 2007) identify principles for identifying and setting priorities for EHS research, including value of information, leveraging research by other governments and the private sector, and adaptive management of nanomaterial EHS research; but it is not clear how these principles were used in selecting the research priorities.

DOES THE STRATEGY IDENTIFY THE RESOURCES
NEEDED TO ACHIEVE STATED GOALS?

The strategy does not identify resources necessary to address questions concerning EHS research needs for understanding nanomaterials and does not identify the projected resources needed to execute the strategy, including funding, education, and training of personnel. This absence of a discussion of resources constitutes a major deficiency. Although the detailed analysis of nanotechnology EHS expenditures in FY 2006 provides information about what was spent during that particular year, there is no assessment of whether the spending was adequate to address EHS research needs voiced by individuals, organizations, and governments worldwide (Denison 2005; Maynard 2007; Ziegler 2007), whether the expenditures by the agencies were appropriate to address EHS research needs based on their missions, or how much additional resources would be required.

From the FY 2006 expenditures, it is difficult even to assess the balance of research among objectives, because in many cases the monetary value of a research project is a function of an agency's budget rather than of scientific needs. However, with respect to the overall funding level, the strategy document suggests that sufficient funding is already being dedicated to EHS research by the NNI and that funds should not be redirected to this research from other kinds of nanotechnology research. The strategy states, surprisingly, that "the current balance of research funding addresses such basic investigations and supports regulatory decision making. Gaps identified in the research that supports regulatory decision making should not be addressed at the cost of broad-based fundamental research—to do so would ultimately undercut the U.S. nanotechnology initiative as a whole" (NEHI 2008, p. 46). An appropriate research strategy should quantify the resources needed to address research priorities, identify where the resources might come from, and ensure that there is adequate training of personnel.

DOES THE STRATEGY PROVIDE ACCOUNTABILITY
FOR ACHIEVING STATED GOALS?

Although lead agencies are identified for each of the five research categories, there is no accountability—no organization or person will be held accountable for the success or failure of the strategy to deliver results. The strategy states that "the success of the strategy . . . depends on the collective efforts of the NNI agencies through their individual and joint activities coordinated by the NEHI Working Group and the NSET Subcommittee. Progress will also depend on the agency priorities and resources" (NEHI 2008, p. 7). That is, accountability is divided among agencies, a working group, and the NSET Subcommittee, and progress depends on individual agency priorities and resources.

In comments to the committee, Terry Medley, of DuPont, stated (Medley 2008, p. 4):

> The executive summary of the document raises two critical questions. 1) Who will implement the strategy? 2) How will the strategy be implemented? With regard to who will implement the strategy, it identifies agencies that will serve as coordinators for the five research areas, it does not explicitly address the coordinating agencies ability to make final decisions regarding the activities in their specific research areas. With regard to how the strategy will be implemented it states that as nanotechnology EHS research and knowledge continue to grow, needs and priorities will evolve. Accordingly, this plan will be reviewed and updated as research progresses. Again, the strategy calls for a coordinated approach as the research progresses, but does not specifically address who has the authority to make changes or revisions needed.

Because of the absence of clearly stated goals and measurable objectives, it is difficult to imagine how the strategy could be used objectively to measure the success of future research efforts. Accountability may require specific quantifiable objectives so that one can determine whether progress is being made.

The strategy does demonstrate how the NNI and other federal agencies have worked together effectively to coordinate their funding and assessment of EHS aspects of nanotechnology and thus avoided, to some extent, unnecessary duplication of research. That is indicative of the function of the NNI, which has been described as a "coordinating platform" (Murashov 2008). However, there is essentially no stakeholder input outside these federal agencies, and in essence the strategy has been constructed in a federal vacuum.

The strategy does not adequately incorporate input from other stakeholders, such as industries that produce nanomaterials and end users of nanomaterials; environmental and consumer advocacy groups; foreign interests, including substantial efforts of other countries; and local and state governments. The committee recognizes that the 2006 and 2007 NNI documents have undergone public comment, but public comment is not the same as actively engaging other stakeholders in the process. In light of the extensive contributions and interests of other nations, in particular the European Union and Japan, it is particularly surprising that the federal strategy appears largely to ignore what other nations are doing. International coordination would help to ensure that there is not unnecessary duplication of research efforts and that data quality is maintained.

To have effective stakeholder engagement requires that the strategy be developed through a process of stakeholder input and consultation. There are many models of this, including tripartite input from government, industry, and civil society representatives, which would ensure that the strategy developed served the needs of regulators, industry, and citizens without being unduly biased by any particular group. Another model is the National Institute for Occupational Safety and Health's National Occupational Research Agenda, in which research needs and directions are developed through a well-established system of stakeholder input (NORA 2008).

Without input from and accountability to external stakeholders, it is not possible for government agencies to develop an effective research strategy to underpin the emergence of safe nanotechnologies. The reason is that federal agencies have a vested interest in justifying the applicability of current efforts rather than critically assessing what is not being done and how deficiencies might be addressed. For example, when agencies are developing their own research strategies, they tend to ask, what research can we do within our existing capabilities?, rather than the more appropriate, What research should we be doing? Other relevant questions need to be addressed, such as, Are resources adequate? Are adequate mechanisms and organizational structures in place to achieve the desired goals? As a result, the federal strategy becomes a justification for current activities based on a retrospective examination that demonstrates success rather than the development of a prospective strategy that questions current practices with an eye to future research needs. That is reflected in remarks by William Gulledge, senior director of the Chemical Products and Technology Division of the American Chemistry Council, an industry trade association, who emphasized the need for a more broadly defined strategy, noting that the NNI plan "'represents a bottom-up approach where agencies identify their priorities. . .We still need a top-down, broad, overall look' at nanomaterials" (Risk Policy Report 2008, p.2).

CONCLUSIONS

The committee concludes that *Strategy for Nanotechnology-Related Environmental, Health, and Safety Research* should not be considered a nano-risk research strategy, because it is missing the necessary elements. Nevertheless, it is important to recognize what the document is and what it has achieved. The NNI strategy represents an impressive collaboration and coordination effort involving 18 federal agencies whose nanotechnology research interests span the gamut from exploratory research (for example, research funded by the National Science Foundation to characterize materials on the surface of nanostructures or nanoparticles) to targeted research (for example, research funded by the Environmental Protection Agency to examine the bioaccumulation of nanomaterials in the food chain). The increased collaboration will probably eliminate unnecessary duplication of research efforts. As the document states, "agencies whose missions support nanomaterial research may use this document to better understand where their activities fit into the overall strategy. Moreover, agencies can use it to identify opportunities for collaboration and cooperation, and manage their relationships with other agencies and their research" (NEHI 2008, p. 6).

The development of the strategy has led to extensive discussion and consensus-building among program managers in the various agencies that participate in the NEHI Working Group and in the NSET Subcommittee; in many cases, these are the same program managers who set priorities and make funding decisions on research proposals (Carim 2008). The strategy is also referenced in requests for research solicitations and has stimulated proposal submissions by individual researchers (Carim 2008). In addition, it has spawned the development of EHS strategies by federal agencies.

The limitations of the document may be due to the NNI-NEHI structure, in that perhaps only a bottoms-up approach could be developed. The NEHI is primarily a coordinating body rather than a visionary one (see Chapter 1). It sees its role as ensuring coordination of activities of otherwise independent agencies that have their own distinct missions. That limits the ability of the NEHI and the NNI to create a vision and an overall plan for federal research to understand potential EHS risks posed by nanomaterials most efficiently. Without an explicit vision or clearly stated purpose, the result of the effort is what is reflected in the document: a compilation of studies rather than a more difficult priority-setting and development of milestones and evaluation measures for determining progress toward a vision. As the strategy states, "development of specific EHS research programs—by NNI agencies singly or jointly—is informed largely, but not exclusively, by the research and information needs of agencies with regulatory and oversight responsibilities" (NEHI 2008, p. 3).

The structure of the NNI and the NEHI, comprising the activities of a large number of diverse agencies with differing missions, makes the development of a visionary and authoritative research strategy extraordinarily difficult. Because the NEHI has essentially no authority over the individual agencies— and so no one agency has authority to shape a research agenda within a second agency—this means that the product of the NEHI can be little more than a compilation of individual agency agendas. Because the NNI has no authority to make budgetary or funding decisions (see Chapter 1) and simply relies on the budgets of its member agencies, it has no resources or influence to shape the overall federal EHS research activity. The NEHI must devise a research strategy that is responsive to individual agency budgetary priorities rather than developing a much-needed vision and strategy that include assurances that adequate resources go to the appropriate agencies to realize the vision. Finally, the NNI has no central figure who is not affiliated with any of the member agencies but is charged with oversight of EHS research and has the budgetary authority to make the necessary research and resource decisions.

Because the NNI is responsible for ensuring U.S. competitiveness through the development of a robust research and development program and ensuring the safe development of nanotechnology, it may be perceived as having a conflict of interest. That may be implied in the previously cited statement in the NNI document that addressing EHS research gaps must not detract from fundamental research to develop the technology. The committee concludes that the conflict constitutes a false dichotomy and that strategic research on potential risks posed by nanotechnology can be an integral and fundamental part of the sustainable development of nanotechnology. Nevertheless, a clear separation of accountability for development of applications and assessment of potential implications of nanotechnology would help to ensure that the public-health mission receives appropriate priority. The nation has addressed concerns about separation of technology development and regulatory oversight authorities for a new and potentially hazardous technology in the past. When both supporters and critics of nuclear energy raised strong concerns about both development and regulatory oversight being housed in the Atomic Energy Commission (AEC), Congress responded in 1974 by creating the Nuclear Regulatory Commission (NRC) to house the oversight function and moved the technology development research into the Department of Energy (U.S. NRC 2008). Congress and the executive branch should consider this model in assuring the safe development of nanotechnology. As an interim step, the NNI Amendments Act of 2008 [H.R.5940.RFS] establishes a separate authority within the NNI with accountability for EHS research.

REFERENCES

Cairns, C. 2008. Presentation at the Second Meeting on Review of the Federal Strategy to Address Environmental, Health, and Safety Research Needs for Engineered Nanoscale Materials, May 5, 2008, Washington, DC.

Carim, A.H. 2008. Presentation at the Second Meeting on Review of the Federal Strategy to Address Environmental, Health, and Safety Research Needs for Engineered Nanoscale Materials, May 5, 2008, Washington, DC.

Denison, R.A. 2005. A Proposal to Increase Federal Funding of Nanotechnology Risk Research to at Least $100 Million Annually. Environmental Defense. April 2005 [online]. Available: http://www.edf.org/documents/4442_100milquestionl.pdf [accessed July 29, 2008].

Kojola, W. 2008. Presentation at the Second Meeting on Review of the Federal Strategy to Address Environmental, Health, and Safety Research Needs for Engineered Nanoscale Materials, May 5, 2008, Washington, DC.

Maynard, A. 2007. Testimony to Committee on Science and Technology, U.S. House of Representatives: Research on Environmental and Safety Impacts of Nanotechnology: Current Status of Planning and Implementation under the National Nanotechnology Initiative, October 31, 2007 [online]. Available: http://science.house.gov/publications/Testimony.aspx?TID=13015 [accessed Aug. 22, 2008].

Medley, T. 2008. Presentation at the Second Meeting on Review of the Federal Strategy to Address Environmental, Health, and Safety Research Needs for Engineered Nanoscale Materials, May 5, 2008, Washington, DC.

Murashov, V. 2008. Presentation at the First Meeting on Review of the Federal Strategy to Address Environmental, Health, and Safety Research Needs for Engineered Nanoscale Materials, March 31, 2008, Washington, DC.

NEHI (Nanotechnology Environmental Health Implications Working Group). 2006. Environmental, Health, and Safety Research Needs for Engineered Nanoscale Materials. Arlington, VA: National Nanotechnology Coordination Office. September 2006 [online]. Available: http://www.nano.gov/NNI_EHS_research_needs.pdf [accessed Aug. 22, 2008].

NEHI (Nanotechnology Environmental Health Implications Working Group). 2007. Prioritization of Environmental, Health, and Safety Research Needs for Engineered

Nanoscale Materials: An Interim Document for Public Comment. Arlington, VA: National Nanotechnology Coordination Office. August 2007 [online]. Available: http://www.nano.gov/Prioritization_EHS_Research_Needs_Engineered_Nanoscale _Materials.pdf [accessed Aug. 22, 2008].

NEHI (Nanotechnology Environmental Health Implications Working Group). 2008. National Nanotechnology Initiative Strategy for Nanotechnology-Related Environmental, Health, and Safety Research. Arlington, VA: National Nanotechnology Coordination Office. February 2008 [online]. Available: http://www.nano.gov/NN I_EHS_Research_Strategy.pdf [accessed Aug. 22, 2008].

NORA (National Occupational Research Agenda). 2008. About NORA…Partnerships, Research and Practice. National Institute for Occupational Safety and Health. Centers for Disease Control and Prevention [online]. Available: http://www.cdc.gov/ niosh/NORA/about.html [accessed May 20, 2008].

Risk Policy Report. 2008. Critics Slam Federal Nano Environmental, Health Research Strategy. Inside EPA's Risk Policy Report 15(20):5-6. May 13, 2008.

U.S. NRC (United States Nuclear Regulatory Commission). 2008. Our history [online]. Available: http://www.nrc.gov/about-nrc/history.html [accessed October 15, 2008].

Ziegler, P.D. 2007. Current Status of Planning and Implementation under the National Nanotechnology Initiative. Statement of the American Chemistry Council at the Hearing on Research On Environmental and Safety Impacts of Nanotechnology, Before the House Committee on Science and Technology, October 31, 2007 [online]. Available: http://www.americanchemistry.com/s_acc/sec_article.asp?CID =655&DID=6334 [accessed Nov. 19, 2008].

4

Review of High-Priority Research Topics, Research Needs, and Gap Analysis

In this chapter, the committee examines the analysis and conclusions presented in Section II (pp. 9-44), "Summary of NNI EHS Research: Portfolio Review and Gap Analysis," of the National Nanotechnology Initiative document *Strategy for Nanotechnology-Related Environmental, Health, and Safety Research* (NEHI 2008). That section discusses research categories, research needs, knowledge gaps, and inventories, and it presents the most specific and detailed technical discussion of topics relevant to decision-making for understanding and assessing the environmental, health, and safety (EHS) implications of nanotechnology. Although the committee perceived the NNI document as falling short of its aim of defining a research strategy, elements of Section II would be important for future development of a federal research strategy.

The committee approached the evaluation of Section II of the NNI document by asking four questions (see Box 4-1) that were directly responsive to the charge to the committee, which was to review the scientific and technical aspects of the draft strategy and comment in general terms on how the strategy would develop information needed to support the EHS risk-assessment and risk-management needs with respect to nanomaterials. The discussion that follows is framed by the preceding materials in Chapters 2 and 3, on the elements of a research strategy, and the committee's own collective assessment of federally funded research in FY2006, which allowed the committee to identify and evaluate the strengths and weaknesses of the NNI document.

As indicated in Chapter 2, an important challenge in developing a risk-research strategy is defining its focus—in effect, the rationale for project selection. Resources are limited, and they must be deployed to create relevant information as efficiently as possible. Embedded in any strategy document are underlying principles that determine the allocation of resources, mechanisms by which research is funded, and how research is evaluated. In connection with the four questions in Box 4-1, those principles determine what is "appropriate" or "correct." The committee believes that the value-of-information (VOI) paradigm

---

BOX 4-1 Questions that Structured the Committee's Analysis

- Is the list of research needs appropriate?
- Is the gap analysis complete and accurate?
- Was the priority-setting of needs correct?
- Does the research support environmental, health, and safety risk assessment and risk management?

---

might have been an excellent approach to informing the development of a research strategy from the outset. The committee recognizes that the 2006 NNI report identified VOI as one of the principles for identifying and setting priorities for EHS research.

A VOI approach would help assess what information would be most valuable in improving understanding of the EHS risks of engineered nanomaterials. Its application relies on assessment of both the quality and the relevance of information, and it necessarily weights efforts in favor of the most pressing research needs.

One fundamental rule of thumb emerging from this approach is that information that cannot change one's (or one's agency's) decision *has no additional value* for decision-making. New knowledge could have other favorable social effects and advance our understanding of the natural world and still not have a place in a nanotechnology EHS research strategy. Application of quantitative VOI approaches clearly is premature, but qualitative concepts could be used in the development of an effective EHS research strategy.

In the review of Section II of the 2008 NNI document, it was apparent that a number of issues cut across most or all of the research priority topics. They are highlighted in the next section of this chapter and are followed by an in-depth technical evaluation of each of the high-priority research topics in Section II that reflects issues specific to the five research categories (Box 4-2). The last section of the chapter discusses the committee's assessment of the current distribution of federal investment in nanotechnology-related EHS research; it became clear to the committee when it evaluated the NNI document that its perception of the balance of relevant research among the five research categories differed substantially from the NNI's perception (see p. 44, NEHI 2008).

CROSS-CUTTING CONCLUSIONS ON ANALYSIS OF
SPECIFIC RESEARCH CATEGORIES

The NNI strategy document organizes EHS research into five overarching topical categories (see Box 4-2), with five research needs in each category. Each category addresses research important to EHS risk assessment. The committee

---

BOX 4-2 Environmental, Health, and Safety Research Categories Identified by the National Nanotechnology Initiative

- Instrumentation, metrology, and analytic methods.
- Nanomaterials and human health.
- Nanomaterials and the environment.
- Human and environmental exposure assessment.
- Risk-management methods.

---

generally agreed that the five categories are logical, complete, and appropriately weighted in scope. The five categories align with the missions and research programs established within and across the regulatory and research agencies that participate in the NEHI Working Group. They provide an excellent organizational framework for describing research activities. Some committee members questioned the position of risk assessment in the document—whether it should be elevated into a separate category or left as an integrating research theme—and this was the subject of some debate. Otherwise, the committee concluded that the basic topics spanned the diverse and complex space of this problem and provided a good organization for the listing of research needs.

The committee found that, with some exceptions, the specific research needs within each category were appropriate for nanotechnology EHS research. The research needs identified substantial aims important for the given research category. However, the committee believed that the lists were incomplete, in some cases missing elements crucial for progress in understanding the EHS implications of nanomaterials or not recognizing common research threads across research categories. For example, the issue of environmental exposure received insufficient emphasis in the exposure-assessment discussion although it was addressed in the nanomaterials in the environment section. The potential for nanomaterials to undergo change within biologic matrices is a common research theme that should be addressed in discussions of nanomaterials and the environment; nanomaterials and human health; and instrumentation, metrology, and analytical methods. Characterization of chemical and biologic reactivity of nanoparticles was not included as a research need in the report. Often, as will become clear, the missing research pieces would have been at an interface between categories, and their absence could have resulted from confusion about where to place them. For example, is environmental exposure a problem best tackled by researchers focused on environmental impact or by those looking at exposure assessment? Missing research needs are detailed in the appropriate sections of the topical reviews that follow.

The gap analysis is neither accurate nor complete in laying a foundation for a research strategy. As discussed in Chapter 3, the NNI strategy document defines a "gap analysis" as a major input in the development of its research strategy (pp. 6-7). The approach of evaluating the status of a specific technical field at a given time (for example, the snapshot) and comparing it with expected or desired goals is a useful exercise. However, the gap analysis by the NNI embodies perhaps the most important flaw that the committee identified in the document. Issues arising from the ineffective gap analysis led to serious deficiencies in all topical categories described in Section II.

The gap analysis was inaccurate because the relevance of existing research projects to the listed research needs was generally overstated. In addition, equating the focus of research projects with research results that address a specific risk-research need is misleading. The document consistently—in every part—assumed that funded projects with only distant links to a research question were indeed meeting that research need. For example, in the measurement and characterization discussion, the development of a subangstrom-resolution microscope was said to fulfill the need "to detect nanomaterials in biological matrices." In another category, human health, it was the committee's expert judgment that more than 50% of the inventoried projects[1] describe research directly relevant to therapeutics rather than to any of the research needs listed as relevant to potential EHS risks related to nanomaterials. The discussion of risk management, for example, considered economists who were collating the anticipated size of the markets for nanotechnology as addressing needs in risk management. The committee considered that many of the 246 research projects listed in Appendix A were of high scientific value but that they were of little or no direct value in reducing the uncertainty faced by stakeholders making decisions about nanotechnology and its EHS risk-management practices. Thus, NNI (NEHI 2008) significantly overestimates the currently funded general research activity focused on EHS research, and this contributes to the inaccuracy of the gap analysis.

The second issue related to the gap analysis is that the approach taken limits the analysis to 1 year (FY 2006) of federally funded research and does not consider EHS research supported by the private sector and elsewhere in the world. Relying solely on U.S. government research has led to a document that lacks the necessary breadth to position our nation's research on the international scene wisely. A recognition of the large-scale effort in Japan (Thomas et al. 2006), for example, to complete exposure and hazard assessments of aerosols might alter the priorities for nanotechnology EHS funding in this country. A more complete gap analysis would cast a far wider net across the technical peer-reviewed literature and related disciplines.

----

[1]The president's 2006 budget considered that there were 43 projects in this category; NNI (NEHI 2008) considered that there were 100 projects, the additional 57 projects being ones that are not "primarily aimed at understanding risks posed by nanomaterials" but also include research on medical-application-oriented research (NEHI 2008; Teague, unpublished material, 2008).

The criteria for priority-setting of research is not clearly stated. Information on priority-setting is only implicit in the graphical timelines (Figures 3, 5, 7, 9, and 11), and rarely explicit in the text. In evaluating each high-priority research need in Section II, the committee consistently observed that there was no clear rationale as to how research priorities were determined. Furthermore, the only representation of research priorities was that implied by the graphical timelines; and the priorities were not discussed at length in the text of NNI (NEHI 2008).

The committee assumes that the criteria for priority-setting stem from NNI (NEHI 2007), *Prioritization of Environmental, Health, and Safety Research Needs for Engineered Nanoscale Materials: An Interim Document for Public Comment,* but that document is cited only once in NNI (NEHI 2008), and then only in the context of establishing the five research categories and 25 research needs. Even if those criteria were the basis of the graphical timelines, the lack of explanation in the text makes it nearly impossible to assess the rationale behind the decisions made by the NNI in constructing the figures. As a consequence, it was generally believed that the absence of more explicit information on priority-setting limits the value and impact of the list of research needs.

In addition, there were a few cases in which the committee questioned the validity of priorities of research needs represented in the graphical timelines. For instance, under research need 2 of the instrumentation, metrology, and analytic methods category ("Understand how chemical and physical modifications affect the properties of nanomaterials," p. 14), it is unclear why "Understanding the effect of surface function on mobility and transformations in water" is considered to have medium-term priority when, given the current production and use of unbound nanoparticles, it must be assumed that nanomaterials are already entering waterways.

The document suffers universally from a lack of coherent and consistent criteria for determining the value of information provided by various research activities and for establishing priorities among the research needs. Criteria and a framework for priority-setting of research would ideally be based on an understanding of the value of each of the research needs and the relationships between them. The committee observed that little or no attempt was made to assess how the information that would be generated by addressing the research needs would be used beneficially. Consequently, there is neither a systematic framework within which research needs can be prioritized, funded, and evaluated nor a mechanism for differentiating between high-cost low-value research and lower-cost higher-value research. Both types of research need to be considered in making pragmatic decisions on directing limited resources to address a specific set of challenges.

For example, many of the research needs and topics listed in the instrumentation, metrology, and analytic methods category are relevant to EHS risk assessment and management, but without a means of distinguishing research with high and low value in addressing potential risks, projects of questionable value are cited as addressing EHS needs. Research listed as relevant to risk in this category includes the National High Magnetic Field Laboratory (National Science Foundation [NSF], project a1-30), Bioabsorbable Membranes for Prevention of Adhesion (National Institutes of Health [NIH], project b2-2), and Using Viral Particles to Detect Cancer (NIH, project b5-6). It is hard to see how such projects will lead directly to information that reduces uncertainty and informs decision-making related to assessing and managing potential risks posed by nanomaterials. If such research is undertaken at the expense of studies of higher value in relation to EHS, it will be indicative of a broken or absent strategy.

A similar situation is found in the Nanomaterials and Human Health research category. In the NNI assessment of relevant FY 2006 research projects, a large portion of the research targets human health through therapeutics. Its primary focus is to develop novel strategies for treating cancer and other ailments that deserve the attention of scientists and clinicians. That may accelerate progress in cancer research and will undoubtedly advance knowledge of nanomaterial-biologic interactions that are relevant to potential risks posed by specific nanomaterials, but it will not contribute directly to the body of knowledge needed to ensure protection of public health and the environment from potential risks posed by nanotechnology and its products. In the detailed assessment of the NNI document that follows, the committee concluded that the current research portfolio does not address the most rudimentary problems in environmental, health, and safety.

ANALYSIS OF SPECIFIC RESEARCH CATEGORIES

The subsections below address the five research categories (see Box 4-2), considering the questions presented in Box 4-1. Each subsection is divided into three parts; the introduction that explains the committee's approach, the evaluation and assessment, and the conclusions.

Instrumentation, Metrology, and Analytic Methods

Introduction

Because the behavior of nanomaterials depends on their structure at the nanoscale (such as physical shape and size and the location and distribution of chemical components), sophisticated characterization and measurement methods are essential for understanding and addressing potential risks.

The potential association between scale-related physicochemical characteristics and biologic effects of nanomaterials challenges conventional approaches to risk. In the past, risk decision-making was typically driven by the chemical constituents of a material, not by physical structure—although there are a few notable exceptions, such as asbestos and the distinctions between inhalable and respirable airborne particles. That approach has generally enabled risks associated with materials to be managed reasonably effectively. But the likelihood that some nanomaterials can cause harm by virtue of their nanoscale structure places a much greater emphasis on aspects of nanomaterials not previously considered important.

The challenges in instrumentation, metrology, and analytic methods for identifying, assessing, and managing nanotechnology EHS effects are threefold: establishing the usefulness of methods currently used to assess risk, translating existing methods to address risk (a process of method bridging), and developing new methods. Those challenges (once risk parameters are clarified) raise three overarching issues: grouping nanomaterials that have similar risk-relevant characteristics, ascertaining the appropriate tolerances of risk-related measurements, and determining the context of risk-related characterization and measurement.

An ability to group nanomaterials according to their biologically relevant behavior is essential if material variants are to be rationalized into a finite number of material classes. Developing methods to assess and to monitor the potential effects of every combination of size, form, chemistry, and other properties of engineered nanomaterials clearly is not feasible. But if materials with similar biologically relevant properties could be grouped, it might be possible to reduce the challenge of characterization to a much smaller set of nanomaterial groups.

Tolerance, the accuracy and precision that measurements need to support risk-based decisions, is likely to vary from nanomaterial to nanomaterial and also over time as new information on the importance (or lack thereof) of specific physicochemical characteristics is developed. Without some idea of the tolerance to which measurements should be made, it is not possible to establish a clear research strategy. For instance, if particles of a nanomaterial have similar biologic behavior whether they are 20 nm or 40 nm in diameter (Jiang et al. 2008b), investing tens of millions of dollars on instrumentation with a resolution of 0.05 nm will not advance their risk assessment and management to any important degree.[2] Understanding appropriate tolerances will be an iterative process that emerges from a well thought-out and integrated research strategy. If resources are to be assigned appropriately, some initial estimates of what is important are needed.

That leads to the third overarching issue: context. Risk-related nanomaterial metrology will depend on the type of material under investigation, the context in which the material is being used (or exposure occurs), and the current level of knowledge on which material characteristics are likely to be important. Metrology requirements for exploratory research on biologic interactions will differ from those for evaluating material toxicity, which in turn will bear only a passing resemblance to measurement and characterization requirements for exposure monitoring and material-dispersion evaluation. Likewise, analytic methods will need to be tied, where possible, to important physicochemical charac-

---

[2]This is a hypothetical example that is loosely based on the Transmission Electron Aberration-Corrected Microscope (TEAM) project discussed in NEHI (2006).

teristics that may differ between nanomaterials. For example, understanding the interactions between gold nanoparticles and DNA will require a detailed understanding of particle shape, size, and surface chemistry; but in monitoring exposure to the same material in the workplace, it may be sufficient to measure mass concentration or surface area concentration for all particles and aggregates that are smaller than a few micrometers in diameter.

In summary, components of an effective research strategy to address nanomaterial instrumentation, metrology, and analytic methods in the context of risk should include

- An assessment of the current state of the art of nanomaterial analysis.
- Classification and grouping of nanomaterials that convey the physical and chemical properties relevant to biologic effects.
- Definition and evaluation of appropriate accuracy and precision (tolerance) for measuring those properties.
- Identification and clarification of the analytic needs of researchers working with nanomaterials in toxicology, exposure assessment, environmental science, and medicine.
- Standardization of methods and metrics used in nanotoxicology studies, including standardized approaches for route of administration and dose metrics.
- Cross-disciplinary translation of established methods to the needs of the nanotechnology-related EHS researchers.
- Development of new methods that meet the specialized demands of nanotechnology-related EHS research.

Evaluation and Assessment

Each of the five identified research needs in this category (NEHI 2008, Figure 3, p. 18) is important for nanoscience and nanotechnology generally (see Box 4-3). However, the breadth of many of the research needs is so great that it is difficult to understand how they will be useful in practice for guiding a nanotechnology-related EHS research strategy.

There is poor balance between near-term needs for research targeted to immediate issues faced by the EHS community (including characterization of nanomaterials in toxicology studies and monitoring of occupational exposures and environmental releases) and evaluation of the efficacy of control and containment measures.

There also appears to be a gap between the identified research needs and the examples of funded research provided in the text that is not clearly resolved (pp. 12-17 and 57-67). Many of the FY 2006 research projects listed in Appendix A as relevant to this research category—although important for the advancement of nanoscience and nanotechnology—have little obvious relevance to EHS issues. There is little effort to address the gap between what is needed and what has been funded.

BOX 4-3 Research Needs for Instrumentation,
Metrology, and Analytical Methods

    1. Develop methods to detect nanomaterials in biological matrices, the environment, and the workplace.
    2. Understand how chemical and physical modifications affect the properties of nanomaterials.
    3. Develop methods for standardizing assessment of particle size, size distribution, shape, structure, and surface area.
    4. Develop certified reference materials for chemical and physical characterization of nanomaterials.
    5. Develop methods to characterize a nanomaterial's spatio-chemical composition, purity, and heterogeneity.

Source: NEHI 2008.

    Research need 1, "Develop methods to detect nanomaterials in biological matrices, the environment, and the workplace," is important but broad and would benefit from being split into three research needs that address biologic matrices, the environment, and the workplace separately. Detecting exogenous nanomaterials in biologic matrices is essential for understanding their movement in the body and doses at the organ, cellular, and subcellular levels. Likewise, detecting nanomaterials in the environment will be essential for both monitoring ecologic exposures and containing possible releases. Workplace exposure is an immediate issue for all of nanotechnology, and methods to address it are necessary. Those three topics underpin much of the research and action needed to understand and address potential environmental and health implications of engineered nanomaterials, and their discussion should be tightly linked to research needs described elsewhere in the document.

    All the specific aims listed under this research need are useful, but they constitute a collection of research interests that lacks coherence. Creating three new research needs would enable more attention to be given to sequencing relevant measurement and characterization research in the context of what is needed to address potential risks.

    In common with other research needs, this section is filled with examples of funded projects that bear little relationship to the overall stated goals. For example, several projects mentioned on p. 13 of the NNI document focus on single-molecule fluorescence. Molecular-level interaction of nanomaterials with cells is interesting, but it does not directly concern detection of nanomaterials in biologic matrices and has little relevance to the practical needs for nanotechnology-related EHS research. Likewise, research aimed at developing nanoparticles as contrast enhancers has limited relevance to the general problem of detecting exogenous nanoparticles within biologic matrices, given that the aim of such research is specifically to develop nanoparticles that are easy to detect. Similar issues arise in the case of cited research on sensors: the projects described are of a general nature, and their specific value to EHS issues is not clear. Without clearer explanation, it is hard to see how, for example, the following projects are justified as addressing nanotechnology-related EHS research needs: National High Magnetic Field Laboratory (NSF, project A1-30), Bioabsorbable Membranes for Prevention of Adhesion (NIH, project B2-3), Using Plasmon Peaks in Electron Energy-Loss Spectroscopy to Determine the Physical and Mechanical Properties of Nanoscale Materials (Department of Energy, project A2-5), and Using Viral Particles to Detect Cancer (NIH, project B5-6).

Research need 2, "Understand how chemical and physical modifications affect the properties of nanomaterals" sits uneasily in this section of the document, as in this area measurement needs cannot be divorced from biological and environmental behavior. It would have been far more effective if research need 2 was directed specifically to issues relevant to biologic and ecologic effects, perhaps by restating it as "biologic" properties. More important, this suggested research need, the correlation of the fundamental structure of a nanomaterial with its biologic properties, does not belong in this research category. Rather, because it is driven primarily by the study of biologic interactions, it should be addressed as a cross-cutting research need between the nanomaterials and human health and the nanomaterials and the environment categories. What does belong in this high-priority group is a discussion of how to characterize the molecular properties of the nanomaterial-biologic and nanomaterial-environmental interface. Information on a nanomaterial's physical and chemical properties is critical for enabling a general understanding of structure-function relationships that will guide future nanotechnology-related EHS research. It is a long-range and exploratory research need, but it is highly relevant to the potential safety or harmfulness of increasingly sophisticated engineered nanomaterials and should form a key component of a strategic research program.

Although the overall need is too broad to be of much use in addressing nanotechnology-related EHS issues, the two specific research subjects identified—"Evaluate solubility in hydrophobic and hydrophilic media as a function of modifications to further modeling of biological uptake" and "Understand the effect of surface function on mobility and transformation in water"—are by contrast too narrowly defined to support strategically relevant progress. These two research areas on their own do not adequately address the studies needed to develop a clearer understanding of how physical and chemical modifications affect the properties of nanomaterials.

Research need 3, "Develop methods for standardizing assessment of particle size, size distribution, shape, structure, and surface area," is based on the fact that such methods are vital for developing a clear understanding of how engineered nanomaterials might affect human health and the environment—and how to avoid the effects. Many of the specific aims listed here are relevant to and important for addressing nanotechnology-related EHS issues. This should remain a high priority research need and receive sufficient attention and support to ensure timely and relevant progress.

What is missing from the strategy document is an assessment of relative importance: What standardization and metrics are suitable for risk assessment and management? Without that context, the research aims become a vehicle to justify broad metrology research across nanotechnology to the detriment of more targeted risk-relevant research. That is especially the case where the precision and accuracy needed for exposure monitoring or toxicity testing are not as high as those needed for quality control or exploratory research.

One emphasis that is essential to this research need but is missing is the importance of community-building activities. Only the broad research community can define and standardize biologically relevant, effective protocols for nanomaterial characterization. The free availability and wide dissemination of methods should be as important an outcome of community-building activities that include round-robin evaluations as the measurement of the accuracy and precision of the methods.

Research need 4, "Develop certified reference materials for chemical and physical characterization of nanomaterials," is important but complex. Standard materials are required to validate the characterization protocols described in research need 3. It is also important to identify metrics with which the standards would be characterized and made available, for example, surface area, size, or chemical activity per unit surface area, such as reactive oxygen species per surface area (Jiang et al. 2008b). Substantial community-building activities (for example, workshops and multistakeholder input) are required to create a pool of useful materials that are relevant to nanotechnology-related EHS research. Efforts to train users to handle and work with the nanomaterials in biologic and environmental testing should also be addressed.

In common with other research needs in the category, the question, How much is enough? is important for assessing and managing risk and is not addressed. Without such understanding of the limitations of reference materials, there are no safeguards to prevent inappropriate levels of investment on irrelevant materials.

Research need 5, "Develop methods to characterize a nanomaterial's spatio-chemical composition, purity, and heterogeneity," is broad, and tolerance and relevance are not addressed in the subtopics. As discussed previously, this research need involves the characterization of nanoscience generally and is ill-suited to the goals of addressing potential EHS effects of nanomaterials. It may be that the intent of this research need was to characterize the nanomaterial-biologic interface. It would be more compelling if it included specific discussion of the critical needs for characterizing this interface and of the tools that could be applied to the needs. Metrology is required that goes beyond nanomaterial detection (research need 1) and nanomaterial gross physical properties (research need 3) because it is important in connection with the molecular-level detail of the nanomaterial-biologic interface. However, to conduct this research requires specific quantitative analysis with the necessary spatial resolution and strategies for handling the challenges of such analysis in relevant biologic matrices.

Some discussion of research in the text (NEHI 2008, p. 16) is not connected to the subtopics in Figure 3. These are important research topics, but their linkages to the identified research needs are not apparent. The descriptions of research projects in the text are generally current exploratory and application-based research projects that in some cases happen to have some relevance to risk. Although the identified research needs and topics intersect to a degree with the needs of the nanotechnology-EHS community (NEHI 2008, Figure 3), the funded programs are often disconnected.

Overall, this section of the report could be improved if it presented a clear strategic route to addressing characterization-related EHS issues. The priorities presented, although reasonable in parts, do not provide such a route.

A notable absence from the instrumentation, metrology, and analytic methods category is research related to the chemical properties of nanomaterials. That would involve adding a topic to research need 5 to address adsorption, compatibility, and reactivity of nanomaterials. For example, the nonspecific fouling of nanomaterial surfaces has important consequences for the absorption, fate, and distribution of the material. Methods to evaluate the corona, the molecules and macromolecules that interact with nanomaterial surfaces, accurately and rapidly are thus of immediate importance. In addition, acellular assays that can monitor reactivity of nanomaterials, such as their participation in the generation and cycling of reactive oxygen species, are important and should be addressed. Another important topic is the change in physical and chemical characteristics of the nanomaterials in biologic systems. For example, nanomaterials of some size may agglomerate to different degrees in a biologic fluid and have different effects (Maynard 2002; Oberdörster et al. 2005; Jiang et al. 2008a).

The 2006 funded projects described in the document do support EHS risk assessment to some extent, but the degree of support is not commensurate with the investment, and the mechanisms to apply many of these research projects to nanotechnology EHS seem to be lacking. The funded projects are important, and they represent a large research investment that broadly advances nanoscience and nanotechnology; but they do not necessarily increase our ability to identify, assess, and manage the potential EHS effects of engineered nanomaterials.

Largely missing are projects that directly advance both immediate applied research and long-range fundamental knowledge specifically directed towards addressing nanotechnology-related risk research.

More effective identification, assessment, and management of nanotechnology-related risk is a challenging goal that will require many resources and focused effort; the current document's description of 2006 research suggests that this investment is not being made.

Conclusions

A strength of this section is that the importance of metrology and analysis is highlighted and recognized. The identification of standard reference materials and methods is notable, and represents some of the research topics (for example, production of commercial samples for workplace monitoring) that need to be present in a federal research strategy. The "Summary of Balance-Assessment for Instrumentation, Metrology, and Analytical Methods Category" (p. 17) is critical for seeing how all the programs fit together; its expansion and a clearer analysis would go a long way toward conveying the big picture.

There is no analysis of the state of the art to justify existing and future research investments. No consideration is given to the relevance of current abilities and methods and the extent to which they negate the need for future research in some fields. For instance, methods already exist to characterize airborne particles by size, mass, surface area, and number concentration that extend down to a few nanometers. Analytic techniques exist that are capable of measuring trace quantities of specific chemicals; and electron and atomic-force microscopy with a resolution of tenths of a nanometer are mature technologies. To what extent are they already being used to address potential nanomaterial effects?

There is no attempt to translate established methods to nanotechnology-related EHS research needs. No consideration is given to how existing and emerging analytic methods might be applied to EHS effects. There is little evidence that characterization techniques in fields outside risk research can be applied to potential effects without substantial investment in translating the technology to a new kind of application or developing risk-specific technologies. Justifying general metrology research as relevant to risk research without appropriate "bridging" is deceptive.

Funded projects are disconnected from research needs. The list of projects funded in FY 2006 and identified as relevant to these research needs seems to be a list of convenience in that it represents current exploratory and application-based research that may have some relevance to addressing risk. Assessing the projects does not provide a strategic route to addressing characterization-related EHS issues. The text is littered with subjective qualifiers: research "can be applied," "could be useful," "will likely benefit." It is the language of wishful thinking, not critical analysis.

Research is not relevant to immediate nanotechnology-related EHS needs. No consideration is given to the accuracy and precision required for risk-relevant nanomaterial characterization. As a result, the research is open-ended and apt to consume considerable resources in addressing questions that are not relevant to protecting public health and the environment. No consideration is given to the different contexts within which risk-related measurements are needed; consequently, there is a danger of substantial research investment in projects and programs that do not address critical issues.

Finally, it is important to keep research on instrumentation, metrology, and analytic methods a primary focus in the federal strategy. Progress in nanotechnology-related EHS research requires advances not just in hazard identification, exposure assessment, standard development, and risk management but in the measurement and characterization of the materials. The current strategy falls short of supporting the necessary research. More effort is needed to ensure that existing and future research efforts address nanotechnology-related EHS needs in a way that provides stakeholders with the knowledge and tools they need to identify, assess, and manage potential risks associated with nanomaterials across their life cycle.

Nanomaterials and Human Health

Introduction

The rapidly expanding development, marketing, and application of nanomaterials with little information on their ability to interact with or disrupt biologic systems raise concerns about their safety in occupational and environmental settings. The safety of nanomaterials is of concern to multiple stakeholders, including government bodies with human or environmental health missions (for example, the Food and Drug Administration, the Environmental Protection Agency [EPA], and the National Institute for Occupational Safety and Health), commercial producers, and nongovernment organizations (for example, the Natural Resources Defense Council, the Environmental Defense Fund, and the American Federation of Labor and Congress of Industrial Organizations). Each of those stakeholder organizations focuses on EHS-related concerns regarding nanomaterials, including medical and therapeutic applications and safety, occupational exposure and worker health, and environmental and consumer exposure and health. The committee reviewed the adequacy of the nanomaterials and human health research section in the context of its completeness, accuracy, and ability to address important EHS issues for each of the stakeholders by addressing the questions posed in Box 4-1.

Evaluation and Assessment

The NNI document identifies five broad, inclusive high-priority research needs related to nanomaterials and human health (NEHI 2008, see Figure 5, p. 24) and specifies a total of 29 focused research topics in connection with them. Each topic is essential for addressing EHS risk assessment and management needs. The emphasis on biologic responses and on exposure routes and measurements is logical and noteworthy. Overall, the list of research needs on nanomaterials and human health is reasonably complete (see Box 4-4). However, some important clarifications and additions need to be incorporated.

The combined list of research in the two needs "Understand the absorption and transport of nanomaterials throughout the human body" and "Understand the relationship between the properties of nanomaterials and uptake via the respiratory or digestive tracts or through the eyes or skin, and assess body burden" is complete except for the absence of an emphasis on quantitative kinetics and application of kinetic models. It should include, where appropriate, quantitative models, such as physiologically based pharmacokinetic models, that account for the influence of physiologic, biologic, and other processes that influence nanomaterial kinetics. Such models would facilitate assessment of interindividual, interspecies, and life-stage-dependent differences in kinetics and dosimetry and other susceptibility factors. The models would constitute a first step in developing more inclusive, integrative computational models for predicting biologic effects.

Moreover, those two research needs describe a single research topic that comprises the human body's absorption, distribution, metabolism and transformation, and elimination (ADME) of nanomaterials. Therefore, they should be integrated into a single research activity focused on understanding the absorption and transport of nanomaterials through the human body and the influence of their physicochemical properties on ADME and toxicity. An important element in understanding the ADME aspects of nanomaterials is to determine whether biologic processes modify the physico-chemical characteristics of the nanomaterial, including changes in surface properties, size, and oxidation state of the components.

The research need "Identify or develop appropriate *in vitro* and *in vivo* assays/models to predict *in vivo* human responses to nanomaterials exposure" emphasizes *in vivo* and *in vitro* hazard-screening tools and offers prediction of biologic response (for example, toxicity) as a goal. The committee notes that prediction of biologic response requires the development of quantitative dose-response data and in some cases mechanistic or mode-of-action data from highly coordinated studies, the articulation of a quantitative representation of the biologic and physical processes, and ultimately the development and use of integrative, quantitative computational (*in silico*) models (ICON 2008). The NNI document should articulate the research required to address each of those steps. For example, both quantitative structure-property-activity relationships and development of biologically based dose-response models should be specifically included as research needs to assist in the integration of data and prediction of toxicity.

The topics identified in the two remaining research needs, "Develop methods to quantify and characterize exposure to nanomaterials and characterize nanomaterials in biological matrices" and "Determine the mechanisms of interaction between nanomaterials and the body at the molecular, cellular, and tissue levels," were deemed complete. However, the committee did consider that,

---

BOX 4-4 Research Needs for Nanomaterials and Human Health

Understand the absorption and transport of nanomaterials throughout the human body.

Develop methods to quantify and characterize exposure to nanomaterials and characterize nanomaterials in biological matrices.

Identify or develop appropriate *in vitro* and *in vivo* assays/models to predict *in vivo* human responses to nanomaterials exposure.

Understand the relationship between the properties of nanomaterials and uptake via the respiratory or digestive tracts or though the eyes or skin, and assess body burden.

Determine the mechanisms of interaction between nanomaterials and the body at the molecular, cellular, and tissular levels.

Source: NEHI 2008.

---

where feasible, it would be prudent to identify activities in each category that complement and influence those in other categories in an effort to promote research coordination. For example, studies addressing research needs in nanomaterials and human health would benefit from a focus on occupationally or environmentally relevant materials, exposure levels, and exposure routes on the basis of well-characterized nanomaterials (research that is addressed in the instrumentation, metrology, and analytic methods category and the human and environmental exposure assessment category). The more integrated approach would increase the value and relevance of the research.

Although the rationale for selecting priorities of the research topics (NEHI 2008, Figure 5) was not clear to the committee, it considered that the sequence of implementation was for the most part logical. An initial focus on development of methods to quantify nanomaterials *in situ* is reasonable because these methods are required for the success of the other research, all of which involves exposure, dose measures, or tracking of nanomaterials in biologic matrices. Initial efforts to identify which portals of entry have high rates of absorption and which organ systems preferentially accumulate nanomaterials were also viewed as appropriate. However, consideration of the diversity of nanomaterials and their applications is critical. Such knowledge should guide selection of appropriate *in vitro* and *in vivo* systems for hazard screening and mechanistic work.

The NNI implied, in Figure 5, that all mechanistic work was of value and should be conducted in the near term. The committee considers that some clarification is needed. Targeted mechanistic research on the interaction of nanomaterials with known biologic or toxicologic pathways and mechanisms (for example, oxidative stress, mutagenesis, or inflammation) addresses important questions about hazard and classification of materials by response in the near term. Although hypothesis-driven, exploratory mechanistic research could address important questions in the near term, purely exploratory mechanistic work might be most valuable if guided by knowledge about relevant exposures routes, end points, and tissues and cell types and may be more useful once some initial research questions are addressed.

The NNI conducted its gap analysis without substantive consideration of the relevance of the research to the two distinct communities that use the information (research focused on clinical uses and patient populations and research focused on occupational and environmental health risks). More than 50% of the projects listed for human health target research directly relevant to therapeutics rather than assessing the potential EHS risks posed by nanomaterials. The committee felt that the relevance of the therapeutic studies was overstated. Three examples of the imbalance in the funded research projects are presented below.

The NNI identified 30 grants as addressing the research need "Understand the absorption and transport of nanomaterials throughout the human body" (which contained seven specific objectives). The sole conclusion regarding gaps identified was that "further research on gastrointestinal and intraocular uptake is needed." However, on closer examination of the 30 grants identified as relevant, only two were focused on issues that directly addressed ADME data that might be useful for environmental and occupational risk evaluation: "Effect of nanoscale materials on biological systems: Relationship between physicochemical properties and toxicological properties" and "Impact of physicochemical properties on skin absorption of manufactured nanomaterials." The remaining 28 funded projects were focused on medical applications of nanotechnology, such as the design of drug-delivery systems or other aspects of therapeutics. Those research projects will undoubtedly generate information that is conceptually useful in understanding the behavior of specific types of nanomaterials, but it is unlikely that they will generate data that would be directly applicable to risk assessment of environmental and occupational health hazards. The NNI document states, with little justification or documentation, that 17 projects directly addressed that research need; this implies that 13 projects have no particular relevance, so it is not clear why the entire budgets of those projects would be included in the tally of funding in this topic. Thus, if one were to carefully examine the FY 2006 funding committed to understanding each of the seven subtopics identified in the research need "Understand the absorption and transport of nanomaterials throughout the human body," there would be at most two grants that might provide useful information in them. It is hard to imagine that the only "gaps" identified by the NNI are in gastrointestinal and ocular uptake, inasmuch as no exposure assessments have been conducted to understand the extent to which gastrointestinal uptake and intraocular uptake are important routes of exposure .

Another example of the flawed gap analysis is in the research need "Identify or develop appropriate *in vitro* and *in vivo* assays/models to predict *in vivo* human responses to nanomaterials exposure." Eight appropriate subcategories were identified (Figure 5). In connection with this research need, only six FY 2006 projects were identified. Although the NNI states that all six directly address the need, examination of their content suggests that only three directly address one or more of the subtopics (B3-1, B3-5, and B3-6 in Appendix A). The other three projects (B3-2, B3-3, and B3-4) may generate relevant information but do not explicitly address any of the subtopics in a way that would be useful for environmental or occupational risk assessment. The NNI acknowledges that the gap analysis is flawed, but it offers no recommendations on how to address this critical limitation. "While there is a low number of projects in this priority research need, this assessment does not capture applicable research in other areas nor many additional research efforts on testing schemes that were not captured by the gap analysis, so a determination of future priorities based on this analysis may be misleading" (p. 22). Indeed, the "Summary of Balance-Assessment" for the section does not mention the paucity of research addressing predictive toxicology for nanomaterials (development and validation of *in vitro* assays that predict *in vivo* toxicity). It is difficult to fathom how two federally funded projects in FY 2006 (B3-1 and B3-5) that directly address the development of *in vitro* and *in vivo* assays and models to predict human response to nanomaterials would be considered a sufficient research effort.

The focus of the research on therapeutics means that the data needs for risk assessment are not being supported. The gap analysis does not accurately or adequately represent research gaps related to nanomaterials that might pose health and safety risks to consumers, researchers, and workers. The committee considers the apparent lack of a sizable number of research projects that directly address the immediate research needs related to potential occupational and consumer risks posed by nanomaterials to be a substantial data gap. Revision of the table (NEHI 2008, p. 20) to separate studies focused on therapeutics from studies that emphasize materials important to these other communities (workers, consumers, and the public) would facilitate a transparent and unbiased assessment of data gaps that will help to spur the needed research.

The small number of projects addressing the research needs in the nanomaterials and human health section and their bias toward therapeutic applications rather than materials relevant to the environmental, occupational, and consumer exposure settings constituted sufficient evidence that the funded research will not support risk-assessment and risk-management needs for these classes of nanomaterials, generate the information needed to support EHS risk assessment and risk management, or provide critical data for regulatory agencies.

Conclusions

There is a need for broad coordination in the parallel pursuit of research needs in the nanomaterial and human health category and across research categories. Research projects in nanomaterials and human health would benefit from research on occupationally or environmentally relevant materials, exposure levels, and exposure routes—work that is carried out in other research categories. A more integrated approach would increase the value and relevance of the research.

The list of high-priority research on nanomaterials and human health is, with few notable exceptions, complete. Additional emphasis of research on the analysis and evaluation of ADME and toxicity of engineered and other nanoscale materials that are related to likely exposures is needed. In particular, there is a need for the collection of quantitative kinetic data and the development of quantitative kinetic models, including, where appropriate, physiologically based models and structure-property-activity models.

The gap analysis was neither accurate nor complete. The gap analysis resulted in the NNI's overstating the relevance of therapeutic studies to the identified research needs and not fairly representing the paucity of projects that truly address the potential EHS risks posed by nanomaterials. Although most of the therapeutic studies are focused on developing novel strategies for treating cancer and other ailments that deserve the attention of scientists and clinicians, they will not directly contribute to the body of knowledge needed to ensure protection of public health and the environment from potential risks posed by nanotechnology and its products.

Nanomaterials and the Environment

Introduction

Nanomaterial exposures and their effects on organisms and ecosystems are influenced by the nature of the material and its applications and will probably depend on the physical and chemical characteristics of the particles, including size, shape, surface chemistry; the frequency, magnitude, and duration of releases or exposures; and countless modifications in material structure and properties mediated by environmental processes. A research strategy that addresses environmental end points must address the breadth of possible variables that may define nanomaterial transport, transformation, bioavailability, bioaccumulation, and trophic transfer and mechanisms that may control toxicity on cellular and organismal scales.

Classically, environmental research has focused on the relationship between chemical composition of contaminants and their environmental behavior and effects. The recognition that nanoscale structure may be more predictive of environmental parameters has forced researchers to rethink concentration-response approaches and place more emphasis on more robust particle characterization in environmental matrices. Broadening exposure characterization will inevitably lead to better predictions of effects.

Several challenges face environmental scientists who are conducting research on nanomaterials: developing reproducible testing methods that provide insight into environmental characteristics, quantifying appropriate effect end points that reflect both physical and chemical stress, developing quantitative structure-activity relationships, and incorporating this information into ecologic risk assessment. Addressing those challenges will require multidisciplinary approaches that include material scientists and physicists in the more traditional environmental collaborations of engineers, chemists, biologists, and toxicologists.

Test methods designed to characterize environmental soluble contaminants may not be appropriate for use with nanoparticles. Quantifying the behavior of a solute in environmental matrices is already challenging; understanding the behavior of nanoparticles may require restructuring assay systems that facilitate particle detection and characterization. That is critical because research has suggested that nanoparticle behavior depends heavily on the characteristics of the environmental matrix. Therefore, it is not sufficient to characterize the test material only before conducting the assay. Particles must also be characterized during the assay, and how their characteristics change must be evaluated (Maynard 2002; Oberdörster et al. 2005; Jiang et al. 2008a). For example, nanoparticle suspensions in freshwater may have aggregation rates that result in substantial changes in aquatic organisms' exposure to them. The response of aquatic organisms may therefore depend on aggregation rate and on exposure duration (for example, continuous vs episodic).

Most ecotoxicologists are not accustomed to quantifying responses of organisms to particles. Although some approaches and insights can be garnered from existing mammal-particle toxicologic research, they will not be useful or predictive for all trophic levels. Research has suggested that aquatic organisms discriminate among colloids of different sizes, but there are no data that support extrapolation of these relationships to nanoparticles (Christaki et al. 1998).

Quantitative structure-activity relationships (QSARs) have been developed for myriad contaminants and used successfully in ecologic risk assessment. Quantifying the influence of nanoscale structure and suspension characteristics (for example, particle size, shape, surface chemistry, and aggregation rate) on environmental characteristics might lead to development of QSAR-like predictive tools. However before such tools can be developed, the appropriate measures of the nanomaterial properties that may affect end points must first be identified, through extensive testing of many different well-characterized nanomaterials for these endpoints.

Current ecologic risk-assessment methods may be a useful starting point, but methods for quantifying nanoparticle-related risk may need to evolve as research on behavior and effects unfolds. It is not apparent that classic metrics for predicting exposure and effects are applicable to nanoparticles. For example, nanoparticle suspensions may have both physical effects associated with their size and shape and chemical effects associated with their surface chemistry and particle composition. The applicability of such measures as volatility or octanol-water partitioning is doubtful.

Elements of an effective research strategy to address the environmental behavior, fate, bioavailability, and effects of nanomaterials and their associated ecologic risk should include

- The development of reproducible testing methods that provide insight into environmental characteristics.
- An assessment of the most important nanostructural characteristics that influence environmental characteristics.
- Determining the appropriate ranges of environmental concentrations to inform effects research.
- Development of mathematical tools that link environmental characteristics to appropriate environmental effects or end points.
- Identification of the appropriate end points.
- Incorporation of nanomaterial research results into ecologic risk assessment and modification of risk-assessment methods to accommodate effects and exposure phenomena peculiar to nanoparticles.

Evaluation and Assessment

The committee reviewed the adequacy of the nanomaterials and the environment section of the 2008 NNI document to assess its ability to encourage research and facilitate quantitative ecologic risk assessment. The strategy was reviewed for its completeness, research priority-setting, and ability to support risk-assessment and risk-management needs.

The NNI document identified five research needs in the category of Nanomaterials and the Environment (NEHI 2008, Figure 7, p. 31). Each of the needs is critical for advancing knowledge and supporting ecologic risk assessment and management (see Box 4-5). The research needs appear to have been derived by extrapolating from the inventory of current research activities rather than as a high-level assessment of near-term to long-term needs. Some discussion of research needs that moves beyond such extrapolation is found in the background paragraphs that describe the need for improved measurement of toxicity, determination of mechanisms of toxicity, development of structure-activity relationships, and consideration of environmental modifications of nanomaterials. Strategic planning for research is also reflected in Figure 7 (NEHI 2008, p. 31). Trophic transfer, including bioaccumulation and bioconcentration, is one possible ecologic end point that, although meriting research, appears to be absent from the proposed strategy. Similarly, it is not clear how weak links in ecosystem-level responses—for example, those related to such ecosystem services as nutrient cycling—will be identified.

BOX 4-5 Research Needs for Nanomaterials and the Environment

　　1.　Understand the effects of engineered nanomaterials in individuals of a species, and applicability of testing schemes to measure effects.
　　2.　Understand environmental exposures through identification of principal sources of exposure and exposure routes.
　　3.　Determine factors affecting the environmental transport of nanomaterials.
　　4.　Understand the transformation of nanomaterials under different environmental conditions.
　　5.　Evaluate abiotic, and ecosystem-wide, effects.

Source: NEHI 2008.

The report sorts 38 projects into the five research needs. The research needs are important for accomplishing the goals laid out for this section. While the goals are presented and described as a priority list, most of the current projects (22 of 38) address research need 3, "Determine factors affecting the environmental transport of nanomaterials." One concern is NNI's priority-setting of research needs. Exposure scenarios should precede toxicity testing for ecosystem risk assessments. Similarly, understanding of environmental fate and transport would be necessary before assessment of organisms at risk. For example, if the behavior of a particular nanomaterial results in sediment deposition, testing effects on sediment-dwelling rather than pelagic organisms might be a priority. While some bioavailability and mechanistic toxicity testing should be a high priority, the committee cautions against extensive toxicity testing without fully understanding environmental fate and transport processes necessary to quantify exposure. Effects characterization without an adequate understanding of environmental exposure may result in resources being expended on research that does not contribute to ecologic risk assessment or facilitate extension to higher-level ecosystem effects. The committee agrees that toxicity bioassay method development must be a high priority.

The first research need concerns the effects of engineered nanomaterials on organisms and the development of methods for measuring the effects at the genomic, molecular, cellular, organismal, and population levels. Determining whether nanomaterials have an effect as defined by a widely accepted, measurable end point is critical for determining whether they should be considered further for the purposes of risk assessment. However, these end points have been developed and refined largely in response to soluble contaminants, not particles. Therefore, the committee supports the priority of research investigations that focus on nontraditional ecotoxicologic end points that are more appropriate for particles, such as nanoparticle effects on protein configuration or phagocytotic responses. Information about testable hypotheses can be gleaned from the scientific literature on the human health effects of exposure to particulate contaminants such as silica, asbestos, and carbon black which have been extensively studied. In addition to different end points, toxicity assessments must include exposure characterizations. The committee supports the priority of understanding the influence of particle characteristics on ecotoxicologic bioassays. Currently these bioassays are always accompanied by quantitative assessments of contaminant exposure (concentration); bioassays of nanoparticles need to include contaminant characterization beyond a mass exposure number. Particle size, shape, surface area, and surface chemistry are all potential determinants in the outcome of biota-nanoparticle interactions. The most important part of this research is the development of sensitive, reproducible ecotoxicologic bioassays for the assessment of the effects of particles.

The second-ranked priority research need is to understand exposure by identifying principal sources and exposure routes. Only one project was identified as addressing that topic during FY 2006. This work should have high priority and should be done quickly because it will inform the array of relevant concentrations to be studied and because it is impossible to predict which organisms will be exposed without adequate exposure characterization. However, the research void introduces considerable uncertainty into the range of concentrations that should be used and even the systems that should be studied. The research void is substantiated by the fact that a search of the EPA Web site yields only one research project focusing on nanoparticle exposure funded in FY 2007 and only three focusing on fate and transport (EPA 2008).

Similarly, in the 2006 inventory, one project was identified as addressing ecosystemwide effects, that is, effects that go beyond those of individual species ("Nanoscale Size Effects on the Biogeochemical Reactivity of Iron Oxides in Active Environmental Nanosystems"). That is not surprising inasmuch as the entire ecotoxicologic literature is slanted to individuals, and few studies focus on higher orders of organization (for example, populations, communities, and ecosystems). The need to cover a wider array of nanomaterials than those of natural origin identified for study in the inventoried 2006 projects is cited in NNI (NEHI 2008). This is a critical need. To apply current knowledge on materials of natural origin to an understanding of risks posed by engineered nanomaterials, more research is needed to understand how the physicochemical properties and toxicity of natural and engineered nanomaterials differ (see discussion of research gaps below).

The largest fraction of research projects on nanomaterials and the environment identified in the FY 2006 inventory investigates factors that affect environmental transport of nanomaterials. The NNI document identifies a lack of emphasis on "more applied" research and little evaluation of existing transport models (NEHI 2008, p. 28). One research need identified here is the determination of physicochemical processes that control the fate and transport of different nanomaterials. Surface modification of nanoparticles in the environment is important because of its potential influence on particle behavior, including agglomeration, aggregation, and sedimentation which may affect bioavailability and possibly nanoparticle reactivity. A more mechanistic approach might provide the foundation of the development of predictive models, provide insights into exposure pathways, and identify organisms at risk. Results of this research may provide insights into exposure pathways and organisms at particular risk, so substantial effort is warranted.

Research need 4 is "Understand the transformation of nanomaterials under different environmental conditions." Physical, chemical, and biologic transformations are all identified as meriting research.

There were 10 projects that were not sorted into the five research needs. Their importance was noted in that they could also lead to nanotechnology applications that contribute to lessening current environmental contamination.

In summary, all the research needs identified as having priority in the NNI document are appropriate and even critical for providing information needed for informed risk assessment. The committee reinforces the need for characterization methods to identify nanomaterials in biologic and environmental matrices and the products of nanomaterial-environment interactions. As stated by the NNI, this must be an overarching consideration. The call to focus on "as-manufactured" nanomaterials may misdirect interim risk assessments by creating large gaps in the understanding of how "manufactured" nanomaterials and those found in natural systems may differ.

With the caveats described above, the priorities in this category are appropriate in that a consideration of hazard below the level of ecosystems often precedes ecosystem-level evaluation. However, estimates of transport and transformation are required to assess environmental exposure and should therefore have higher priority than evaluation of ecosystemwide effects because the latter cannot be usefully studied without knowing what the likely environmental concentrations will be and what organisms might be exposed. Therefore, the committee recommends that the research needs be rearranged as (2), (4), (1), (3), (5). Exposure and transport processes would be characterized before effects. That would provide a rationale for the selection of bioassay species. Transformation processes would be characterized before higher-level ecosystem effects. At present, the distribution of projects among the research needs does not appear to be consistent with the proposed priorities or with our recommended sequence. Attention should be given to making resource allocation consistent with the prioritized research needs.

Although the research strategy appears to reflect an important collection of existing federally funded research, there are several gaps in the identified research needs:

• The strategy document does not specifically identify the need for studying naturally occurring or incidental nanoparticles that have similar structures or that may be identical with manufactured nanomaterials.

• The document does not identify development of protocols to evaluate nanomaterial loss from products as a research need despite an apparent trend toward using nanomaterials predominantly as composites in more complex matrices of resins, fabrics, and coatings.

- The document does not consider characterization of bioavailability and toxicity of nanoparticles in complex media, such as effluents. It is important because many nanoparticles will enter the environment in effluents and discharges.

- The document does not mention the need to characterize interactions among nanoparticles and other environmental contaminants. Such interactions could alter environmental behavior, bioavailability, and toxicity of nanomaterials.

- Characterization of nanoparticle transport through food webs is critical for ecosystem health, including potential human exposure.

- Methods for identifying nanomaterial sources, such as isotopic "fingerprinting" techniques, and modeling techniques to track movement of nanoparticles in the environment are needed.

- Research to assess the potential environmental "collateral damage" associated with nanomaterial fabrication needs to be clearly linked to life-cycle analysis mentioned in the NNI document.

The latter topic goes far beyond using off-the-shelf technologies for risk management in material production. It requires an assessment of the quantities and qualities of wastes generated in manufacturing specific nanomaterials and of the risks associated with handling and disposing of the wastes and of the feed stocks used in the manufacture of nanomaterials.

Although the document notes the need to develop methods for characterizing nanomaterials in complex matrices, it does not describe a mechanism for ensuring translation of method developments that may occur in the biomedical sciences, fundamental nanochemistry, or elsewhere in the EHS community. The disconnect between ecologic risk-assessment and risk-management methods for particles vs solutes has not been addressed, and environmental scientists are left to borrow from the human health literature on particles. Although much can be learned from the extensive literature on the impact of particles on human health, caution is needed when making extrapolations to ecologic endpoints, because of the potential differences in exposure scenarios and in physiology and biochemistry among organisms. Development of ecologic risk-assessment and risk-management tools should progress in tandem with the research on fate, behavior, and toxicity already identified.

Conclusions

A strength of this section is that the major topics identified for research are appropriate. Each is critical for meeting the ultimate goal of risk assessment and material management.

Several important research topics have been overlooked. It is important that the research strategy be comprehensive so that high-priority research can be accomplished in a logical manner. Research needs must be comprehensive to ensure that ecologic risks can be assessed and nanomaterials managed objectively and with minimal uncertainties.

There was no justification for the setting of priorities of the research needs, nor were they set in relation to resource allocation. The priorities of research needs were not well justified, and even a cursory examination suggests that a different prioritization might be more logical. Projects funded in FY 2006 and identified as relevant to the research needs do not support the proposed prioritization.

Priority of research on factors that control transport, fate, and exposure should be expressed in a fashion that clarifies the need for this work to inform ecotoxicity studies. This is a critical inaccuracy in the document. The document suggests that ecotoxicity research should proceed immediately without attention to identifying species at risk on the basis of an understanding of nanoparticle behavior, fate, and transport. That could result in a substantial waste of resources.

Human and Environmental Exposure Assessment

Introduction

For nanomaterials to present a risk to human health or ecosystems, both exposure and hazard must exist. Without knowledge about exposure potential at some point in the life cycle of nanomaterials, it is not possible to assess risk appropriately or to implement well-founded risk-management practices. Research conducted with the goal of assessing potential exposure to nanomaterials must take into account the physicochemical properties of the nanomaterials because they affect partitioning from portal of entry to secondary compartments in the human body and the environment. The risk-assessment paradigm (NRC 1983) connects exposure to dose to response. This section focuses primarily on exposure and dose. Dose-response relationships are addressed in other sections of the report.

One of the strengths of the 2008 NNI strategy document is that it clearly identifies exposure research as a high-priority need and articulates its relevance to risk assessment. It also highlights the paucity of research in this regard and reflects on the nascent nature of nanotechnology (NEHI 2008, p. 34) and lack of exposure information.

Because exposure is a critical determinant of dose, exposure-assessment information will be necessary for informing the design of toxicologic and ecotoxicologic studies with respect to exposure in animal and *in vitro* studies.

But the exposure-dose relationship needs to be considered critically in assessing nanomaterial interactions with organisms and the environment. For example, most of the studies on the assessment of toxicity of nanomaterials have used extremely high exposure concentrations (doses), which are usually irrelevant in realistic exposure scenarios (Oberdörster et al. 2005) except possibly industrial exposures and accidents. Although such high-dose studies can identify a hazard, they also lead to identification of mechanisms that may not be relevant at lower exposures and thus may contribute to an unrealistic perception of risk. In addition, most of the studies have focused on acute exposures and neglected chronic and environmentally relevant exposures.

Evaluation and Assessment

The NNI document identified five research needs in the category of Human and Environmental Exposure Assessment (NEHI 2008, Figure 9, p. 36). The five research needs (see Box 4-6) are all important, but they are not well elaborated. As an organizing principle, the NNI document (p. 33) adopts the approach of identifying and characterizing exposed populations by categories and relating their exposures. The committee believes that the broader concept of human and ecologic exposure potential throughout the life cycle of nanomaterials (from manufacture to packaging, distribution, consumer use, and disposal) needs to be considered as an overarching research theme. In addition, with respect to human exposures, the document focuses mainly on occupational issues. Environmental exposures receive little attention in this section except as conceptualized in Figure 10 and wording in section III (p. 46) that calls out the need to characterize the health of and presumably identify exposures to environments. Issues related to environmental exposure are also addressed briefly in the category "Nanomaterials in the Environment."

BOX 4-6 Research Needs for Human and
Environmental Exposure Assessment

1. Characterize exposure among workers.
2. Identify population groups and environments exposed to engineered Nanoscale materials.
3. Characterize exposure to the general population from industrial processes and industrial and consumer products containing nanomaterials.
4. Characterize health of exposed populations and environments.
5. Understand workplace processes and factors that determine exposure to nanomaterials.

Source: NEHI 2008.

The gap analysis presented in the document lacks substantive discussion of exposure except for a cursory treatment of occupational exposure. The committee noted that the NNI did not identify the lack of research on exposure throughout the life cycle of nanomaterials as an important gap. That omission appears to be due to the lack of research projects on this subject in the portfolio of FY2006 projects.

Understanding metrology and developing tools to characterize and measure attributes of nanomaterials—including particle size, number, and surface area—relevant to exposure is not identified as a research need, and it is implied that it is adequately addressed by a few projects in the instrumentation and metrology section (p. 33). Of particular concern is the challenge of assessing "dose" in toxicologically relevant terms. Although this is not a new challenge in the field of toxicology (appropriate dose metrics for particulate matter exposure have been studied for decades), whether nanomaterial "dose" is best assessed by particle mass concentration, surface area, concentration of reactive functional groups, or other means, will be an especially important area for standardization in nanotoxicology research.

Types of research that should be considered include the following:

- Developing instrumentation for personal monitoring.
- Monitoring air and water discharges in the workplace.
- Research on exposure associated with product use throughout the life cycle from manufacture to distribution and consumer use to disassembly and disposal (Thomas and Sayre 2005; Borm et al. 2006).
- Research on source apportionment, for example, exposure to materials of manufactured origin relative to exposure from naturally occurring or non-manufactured anthropogenic materials, such as combustion products.
- Research on contributions of specific nanoparticles to total exposure, including personal exposure (personal samplers) vs area exposure.
- Research on personal susceptibility because lessons learned from exposure to particulate matter (including ultrafines) suggest that such factors as age, sex, windows of exposure, genetic makeup, and pre-existing diseases can play a critical role in susceptibility.
- Research on routes of environmental exposure, including commercial trends and the potential for nanomaterial penetration into conventional material markets, with an assessment of the unintended and associated environmental losses.
- Development of methods of identifying environmental "hot spots," including fundamental studies of nanoparticle movement through the environment and interactions with known environmental pollutants.
- Research on trend forecasting, using tools from social sciences to allow gross exposure assessment and more targeted studies. Some nanomaterials have been produced and used for decades in large quantities, such as TiO_2 and carbon

black (although these are not "engineered"); in the case of carbon black in particular, several epidemiologic studies have begun to capture workplace exposures (Morfeld and McCunney 2007).

The ordering of research needs in exposure research appears incongruous. For example, although characterizing workplace exposure appears to have the highest priority for research in this category, it seems misplaced with respect to research need 2, which aims to identify population groups and environments that may be exposed to engineered nanoscale materials. Similarly, research need 5 seems to be required to arrive at the conclusion that characterization of work place exposure should be important for research. Indeed, understanding which population groups and environments may be exposed appears to be a prerequisite for selecting the type of workplace settings that should be the focus of research to characterize exposures among workers. Both research needs 1 and 5 appear to have been eliminated in the final list in Section III (p. 46). In general, research priorities seem to have been simply an articulation of the collection of existing research in FY 2006, not priorities for research required to address knowledge gaps. Appropriate priority-setting of research would enable proper allocation of resources. That does not necessarily imply a chronology of research; many types of important research can and should be addressed in parallel.

As presented, there will be large gaps in exposure-assessment information needed for EHS risk assessment and management. There appears to be a lack of clarity as to how and where exposure issues need to be addressed. They are scattered among several sections of the document with no apparent linkage. And the critical linkage between environmental and human exposure is overlooked. Because ecologic exposures may be more difficult to assess than occupational exposures because there are more uncontrolled variables, it is important that environmental exposure research be a priority, and greater recognition of the commonalities of this research need to both the Human and Environmental Exposure Assessment and the Nanomaterials in the Environment categories is needed.

The research priorities described in the NNI document will potentially support environmental health and safety research needs, but they are largely insufficient to allow for rigorous exposure assessment. Information on exposure to engineered, incidental, and natural nanoparticles is critical for development and implementation of effective risk-management plans.

Conclusions

The NNI acknowledges the importance of exposure research (primarily in occupational settings), but the research portfolio, gap analysis, and priority order do not adequately reflect attention to it.

The 2008 NNI document does not address human and environmental exposure potential throughout the life cycle of nanomaterials. It focuses primarily on occupational exposure.

The exposure-assessment section is imbalanced and does not adequately connect with research on environmental processes that determine environmental exposures.

Understanding metrology and developing tools to characterize and measure attributes of nanomaterials—including particle size, number, and surface area—relevant to exposure is not identified as having high priority, and it is implied that it is adequately addressed by the projects listed in the instrumentation and metrology section.

The document does not consider exposure in the context of susceptible populations in humans and the environment, nor does it consider the need to identify such populations. An exposure that may be harmless for a healthy organism may be detrimental to a susceptible population.

The NNI document does not address the importance of exposure studies in the design of toxicologic and ecotoxicologic studies. Repeat or chronic studies in relevant experimental animal models and model systems using realistic exposure concentrations should be an essential component of risk assessment of nanomaterials (including considerations of susceptibility, mechanisms, and mode of action).

Risk-Management Methods

Introduction

By including risk-management methods as one of its five research categories, the 2008 NNI document recognizes that research on risk management can not only broaden available options but also inform risk-assessment research. For an emerging set of technologies, such as nanotechnology, with great uncertainties regarding hazards and exposures, the rapid and active development of risk-related information for risk management should have very high initial priority.

The NNI document identifies five research needs (see NEHI 2008, Figure 11, p. 42 and Box 4-7) that, with several exceptions, subsume the twenty-four research needs in NEHI (2006). There is no description of the process by which these changes occurred. NEHI (2007) provides a limited description of the combining and prioritization of the 2006 research needs, but does not account for why some identified needs (for example, packaging needs, spill containment methods) are not mentioned. In addition, many of the specific research needs subsumed under the five research needs in NEHI (2008) are only evident in the report's Figure 11 and are not discussed in the text.

Responsible nanotechnology-related risk management requires not only research to support risk assessment and to develop new knowledge about risk-management methods and technologies but data collection on trends and practices and dissemination of risk information. A research strategy for risk-management methods should lay out clearly the boundaries between research activities and risk-management data-collection activities. Those boundaries are not defined in the 2008 NNI document. Instead, some essential data-collection and information-dissemination activities are listed as research projects. Such activities are critical for effective risk management, but they do not constitute risk-management research. For example, collecting information on nanoparticle type, composition, and physicochemical characteristics is not research; development of a control banding method[3] based on those characteristics would be.

Evaluation and Assessment

The NNI document lacks a rationale for the selection of research needs and assignment of specific projects related to risk-management methods. That is evident from the statement on p. 41 that indicates that this category has been used as a catchall for projects otherwise not classifiable: "issues not typically thought of as pertaining directly to risk management needs, such as ethics and societal considerations, are included in the projects that fall under this category." Nearly half the already small number of projects, and 62% of the total funding, could not be assigned to any of the other four categories so were placed here. The text does not describe how the unclassifiable projects contribute to meeting research needs.

Ideally, the NNI and the Nanotechnology Environmental and Health Implications Working Group (NEHI) would constitute a useful structure for bringing the needs of risk managers in the regulatory agencies to the attention of scientists in the primary research agencies. The NNI strategy states that "input about the needs of regulatory decision makers expedites the development of information to support both risk assessment and risk management of nanomaterials" (p. 3). That might be true, but there is no description of input from agency risk managers in the 2008 NNI document. Moreover, this section addresses only occupational settings; risk managers for the Food and Drug Administration and EPA would most likely have included environmental and consumer exposure settings as well. The focus of the research may be partly due to NNI's own data collection methods, as NNI acknowledges on p. 38, "the apparent lack of fund-

---

[3]"Control banding is a qualitative risk-assessment and risk-management approach to promoting occupational health and safety." For additional information, see NIOSH (2005).

BOX 4-7 Research Needs for Risk Management Methods

    1. Understand and develop best workplace practices, processes, and environmental exposure controls.
    2. Examine product or material life cycle to inform risk reduction decisions.
    3. Develop risk characterization information to determine and classify nanomaterials based on physical or chemical properties.
    4. Develop nanomaterial-use and safety incident trend information to help focus risk management efforts.
    5. Develop specific risk communication approaches and materials.

Source: NEHI 2008.

---

ing by regulatory agencies for risk management methods research could be due to the data call having been focused primarily at grant-related efforts for a topic that may not always be addressed through research."

There is very little indication of priorities among research needs in this section. Most of the text describes the existing studies that have been placed in this category and the substantial gaps in most of the research needs. There is no textual description of priorities among the many gaps or of how the gaps will be strategically filled.

The only indication of priority among the research needs is in Figure 11. Of the 13 subjects in the five research needs, all but two indicate high priority for immediate emphasis. That is appropriate for risk management of an emerging technology, but it is not informative, especially given the poor description of what is involved in the research needs. Moreover, in a research field characterized by uncertain risks and poor-quality information about risks, it is not appropriate to stall the development of essential risk communication, but this is the only research need that is put off to the intermediate term.

In reviewing this research category, the committee compared the description of research and research needs in risk-management methods in the 2006 NNI report with the research needs, listed projects, and text discussion on risk-management research in the 2008 NNI document. Research gaps were identified through the comparison and with expert judgment, and the evaluation of priorities was based on the descriptions in the 2008 document. Because the content is explicitly related to risk management, the question of relevance to risk management was not considered separately.

Analysis of Individual Risk-Management Research Subjects

The strategy briefly describes 14 projects in the risk-management methods research category, with a total funding of $3.3 million, primarily from NSF and the National Institute for Occupational Safety and Health.

In many cases, it is difficult to discern from the information provided in the 2008 NNI document what is intended by the category; this complicates an independent analysis of the appropriateness of the research needs. For example, research need 3, "Develop risk characterization information to determine and classify nanomaterials based on physical or chemical properties," implies development of a banding or other screening-level categorization of nanomaterials for risk-management purposes on the basis of readily available physical or chemical characteristics. That is a highly relevant and appropriate research need for risk management that is referred to in the 2006 NNI report. The 2008 document, however, does not describe the research need in any detail or how it is to be met. The text combines the research need with the unrelated research need 4, "Develop nanomaterial-use and safety-incident trend information to help focus risk management efforts," apparently because one 2006 project was believed to address the two rather disparate research subjects equally. In place of a thorough description of the research needs, the text describes the severe limitations of the one project placed in this grouping.

The discussion of research need 3 (risk-characterization information) and research need 4 (trend information) also illustrates the failure of the section to distinguish between risk-management method research and risk-management activities. Compiling information on use, trends, and products is essential for developing appropriate risk-management strategies. However, it is not clear why developing a Web-based library (research need 3, project E3-1 in Appendix A, p. 87) or collection of trend information (research need 4) is considered as filling a "research need" instead of as an infrastructure or surveillance activity, especially when it is only a voluntary activity and therefore unlikely to be comprehensive or representative in its characterization. Moreover, the information collected is stated to be "nanomaterial-characterization" rather than "risk-characterization" information identified as a research need. That is another example of how the document is compromised by its efforts to make existing projects fit into the research needs previously identified as critical even when the projects are neither truly research projects nor designed to develop information pertinent to the research need.

Research need 1 (workplace practices and environmental controls) has a primary focus on inhalation exposure; only respirators and personal protective equipment are mentioned. Projects assigned to this research need were relevant and designed to provide essential information. The committee notes, however, that studies of workplace design and other engineering controls, dermal and other routes of exposure, and workplace hygiene and disposal practices should also be discussed in the section. There are large gaps in worker-protection research, and little in this document indicates strategies or priorities for filling them.

Research need 2 deals with life-cycle analysis and comprehensively considers, "manufacturing, incorporation into an integrated product, consumer use, and recycling or disposal" (p. 4). It is essential that not only the finished product but the materials, byproducts, and waste in producing the materials be considered with regard to EHS. But the description of this research need does little to explain the strategic approach to understanding product or material life cycles. The 2006 portfolio identified only two projects in this category, one of which is a life-cycle analysis of manufacturing technologies rather than products or materials (project E2-2 in Appendix A, p. 87); the other is limited to a small sector of products (project E2-1). The strategy itself identifies a clear research gap in life-cycle analysis for product classes not considered in the two current research projects. The document suggests that the research gap is so large, "a systematic evaluation . . . is needed to evaluate where the most critical of such gaps would exist" (p. 40). However, there is no further discussion of conducting such an evaluation. Thus, although including life-cycle analysis is appropriate, a clearer description of specific research and of how the extensive gaps are to be filled is needed.

Only one project is identified in research need 5 (risk-communication approaches). It is restricted to workplace-related issues, and this indicates a large gap in risk-communication approaches for the general public. In addition, the single project listed describes an information-dissemination project rather than a two-way risk-communication project. The document should consider risk communication as a useful information-gathering process and give higher priority to problem scoping and formulation processes with interested and affected parties (NRC 1996).

The section on risk-management methods identifies four gaps on p. 41 of NNI (NEHI 2008): trend information, exposure controls, flammability or reactivity changes due to particle size, and material-safety data sheets. In the broader summary of research needs on p. 46, the 2008 NNI document identifies three major risk-management research gaps to be addressed in the near term: "develop risk characterization information to determine and classify nanomaterials based on physical or chemical properties," "develop nanomaterial-use and safety-incident trend information," and "expand exposure route-specific risk management methods research and life cycle analysis research on the basis of nanomaterial use scenarios expected to present greatest exposure and potential for health or environmental effects." The committee agrees that these seven research priorities, some of which are identical with the research needs mentioned in the document and some not, are reasonable. The lack of concordance between the two lists of identified gaps, however, and the lack of discussion of how the NNI and the NEHI intend to promote research to address them preclude useful evaluation of whether the NNI document provides a useful strategy for filling gaps and meeting short-term and long-term risk-management needs.

Risk-management topics and kinds of research areas in addition to the gaps identified by the document should be considered in this section. They include identifying nanotechnology-enabled products that can assist in managing risks posed by conventional hazards, and permitting the replacement of hazardous chemicals with less hazardous materials. For example, the document indicates that the properties of nanomaterials can be used to "clean contaminated soil and groundwater" (p. 3). That suggests an important risk-management activity for EPA. Although this kind of research was mentioned in the 2006 NNI report and research project C4-8 in Appendix A (p. 82) appears to support it, there is no further discussion of it in the 2008 document. Identifying and developing nanotechnology-enabled risk-management approaches to environmental problems should be addressed as a separate research need.

Conclusions

The criteria for setting priorities for risk-management methods research were not clearly stated. Information was only implicit in the graphical timelines, not described explicitly in the text. Descriptions of high-priority research needs and how they are to be met are lacking; in their place are descriptions of the FY 2006 projects and their limitations in meeting the needs. There is inadequate description of the process by which the 24 research needs identified in the 2006 NNI report were culled to the five in the 2008 NNI document. The graphical timeline gives high priority to nearly all research needs, providing little strategic guidance for meeting them within resource constraints.

The gap analysis for risk-management methods is flawed and limited by the decision to use the 2006 research portfolio as its basis. Major gaps, including management of environmental and consumer risks with emphasis on potential risks to infants and children, are not addressed. The small number of research projects in this category and the smaller number of research projects that actually address the identified research needs underscore the enormous gaps between what is needed and what the agencies are doing. The failure to distinguish carefully between risk-management methods research and risk-management data-collection activities further hampered the gap analysis. The lack of consideration of management of environmental and consumer risks constitutes another considerable gap. It pertains to consideration of risk-management approaches to both general population exposures and specific potential exposure settings, such as accidents and spills, environmental discharges, and exposure through consumer products with the likelihood of exposure of infants and children; it also pertains to the development of life-cycle analyses, which must encompass not just manufacturing processes but the entire product life cycle from resource extraction through disposal. In general, approaches to risk management, such as control banding, that can help to address risks in the absence of completed traditional risk assessments are not adequately addressed in the document. Although the focus on workplace risk management is reasonable given that the occupational setting is likely to be the initial setting where important exposures occur, and the few projects that assess the adequacy of exposure-control measures are critical and appropriate, the overall risk-management research portfolio and strategy are inadequate to address societal needs.

The document does not provide evidence of a strategic approach to risk-management research. The need for the rapid development and validation of effective risk-management methods is great for a set of rapidly emerging technologies like nanotechnology, but the narrow focus on 2006 studies and failure to describe adequately what is meant by the research categories and how projects are to be given priority constitute a failure to develop a strategic plan to meet the need.

COMMITTEE'S ASSESSMENT OF CURRENT DISTRIBUTION OF FEDERAL INVESTMENT IN NANOTECHNOLOGY-RELATED ENVIRONMENTAL, HEALTH, AND SAFETY RESEARCH

The NNI comments on the distribution of nanotechnology-related EHS research investment by illustrating the amount of money it was spending on each of the five research categories in FY 2006 (see Table 4-1). It states that "it is appropriate that investments at this time are predominantly in the categories of Instrumentation, Metrology, and Analytical Methods, Nanomaterials and Human Health, and Nanomaterials and the Environment. The balance of spending will evolve in time as research programs mature and efforts that are undertaken sequentially are initiated" (p. 44).

On the basis of the breakdown in funding, the NNI concludes that, "in short, the analysis demonstrated that the Federal Government is supporting more EHS research than has been previously identified, and the research is well-distributed across key priority areas" (p. 2). However, the analysis does not address how well the funded studies are addressing the specific research needs for a science-based assessment of the human health and environmental risks posed by the production, use, and distribution of nanoscale engineered materials. In the committee's opinion, examining what is funded (Appendix A, pp. 55-58) leads to a different research portfolio that is heavily slanted to specific medical-imaging applications, therapeutic nanomaterials, and targeted drug delivery, especially cancer chemotherapeutics, and to studies focused on understanding fundamentals of nanoscience that are not explicitly associated with the EHS aspects of the risks posed by nanomaterials.

The nanomedicine projects are not basic toxicologic studies of potential human response to nanomaterials in general. Rather, much of this research focuses on finding new applications of nanotechnology-related therapeutics. That does not lead to the general understanding of factors governing absorption, distribution, metabolism, elimination, and toxicity of manufactured nanomaterials needed for a comprehensive risk assessment of manufactured nanomaterials with respect to environmental, occupational, and consumer exposure (for example, cosmetics).

TABLE 4-1 NNI Evaluation of Federal Grant Awards in FY 2006 That Are Directly Relevant to EHS Issues

Category	Number of Projects	$ Invested (Millions), FY 2006
Instrumentation, Metrology, and Analytical Methods	78	26.6
Human Health	100	24.1
Environment	49	12.7
Human and Environmental Exposure Assessment	5	1.1
Risk Management Methods	14	3.3
TOTAL	246	67.8

Source: NEHI 2008.

Many of the funded projects will not generate the information needed to support EHS risk assessment and risk management or provide critical data for regulatory agencies. It makes no sense to include many of the projects listed in Appendix A only because incidental knowledge, procedures, or techniques obtained from that research might be relevant to one or another aspect of research relevant to EHS needs in nanotechnology. The committee notes that the NNI chose to include an additional 116 projects in Appendix A that were not included in the president's budget even though they were aimed primarily at medical applications or at characterization and measurement of nanomaterials (NEHI 2008; Teague, unpublished material, 2008).

The committee conducted its own informal reassessment of the current balance of nanotechnology-related EHS-research investment by using its professional judgment. The committee reviewed the titles and abstracts of the projects to determine which are *primarily* aimed at understanding the potential risks posed by engineered nanomaterials or would otherwise be reasonably expected to provide data that are directly relevant to EHS evaluation. The results are presented in Table 4-2. (Only the percentages of projects in each broad category are presented, because the funding of each project was not readily available.)

Table 4-2 shows that roughly one-fifth to two-fifths of research projects in the instrumentation, metrology, and analytic methods category and about one-third of projects in the human-health category are directly relevant to understanding the potential risks posed by engineered nanomaterials or would otherwise be reasonably expected to provide data that are directly relevant to EHS evaluation. The ranges in Table 4-2 reflect the variability in professional judgment among committee members; such an evaluation has elements of subjectivity. Nevertheless, what is critical is that fewer than half the projects listed in Appendix A are relevant to understanding of EHS issues related to nanomaterials. Therefore, the amount of money being spent by the federal government specifically to address EHS needs in nanotechnology is certainly far less than the

TABLE 4-2 NRC Committee's Estimate of Percentage of FY 2006 Projects That Are Aimed Primarily at Understanding Potential Risks Posed by Engineered Nanomaterials

Category	Committee's Professional Judgment
Instrumentation, Metrology, and Analytical Methods	18-40%
Human Health	30-32%
Environment	67-84%
Human and Environmental Exposure Assessment	100%
Risk Management Methods	57-78%
TOTAL	36-48%

$68 million indicated in the NNI strategy document. It should be noted that that conclusion is supported by other independent analyses of the issue (for example, GAO 2008; Maynard 2008).

CONCLUSIONS

Cross-cutting observations that are relevant to all research categories in the 2008 NNI strategy document include the following: generally appropriate research needs are identified, priorities among research needs are not clearly articulated, and the gap analysis contributes to overstating the amount of relevant federal research being conducted to support EHS research needs related to nanomaterials.

The organization of research into five topical categories is necessary, but it obscures the interrelationships among research needs and creates the possibility that research needs that fall between categories will be overlooked. It is important that the research categories not be viewed as silos. For example, environmental exposures is a common thread in both research categories; Nanomaterials and the Environment and Human and Environmental Exposure Assessment. An example of a research need that may have been omitted because it falls between categories is the omission of characterization methods that consider specific biologic settings. Additional examples are discussed in Section II.

Inventories of the research needs are sufficient for some topical categories, but they are poorly defined and incomplete in risk management and exposure assessment. For example, the discussion of exposure assessment does not address exposures throughout the life cycle of nanomaterials and the discussion of risk-management methods does not cover management of environmental and consumer risks, including specific potential exposure scenarios, such as accidents and spills, environmental discharges, and exposure through consumer products.

Poor gap analysis is a problem in all sections of the document, but it is particularly severe in the discussions of human health and metrology. Table 4-2 offers the committee's collective expert judgment of the extent to which the NNI strategy document miscounts research projects in its gap analysis. As is apparent, this problem was particularly severe with respect to the instrumentation, metrology, and analytic methods category and the human-health category. The extent of the problem is so great that the committee is concerned that the current funding or allocation of funding for EHS research needs related to nanomaterials may not be adequate to address current uncertainties in the manner needed to understand the risks posed by nanomaterials.

REFERENCES

Borm, P.J., D. Robbins, S. Haubold, T. Kuhlbusch, H. Fissan, K. Donaldson, R. Schins, V. Stone, W. Kreyling, J. Lademann, J. Krutmann, D. Warheit, and E. Oberdorster. 2006. The potential risks of nanomaterials: A review carried out for ECETOC. Part Fibre Toxicol. 3(1):11.

Christaki, U., J.R. Dolan, S.P. Pelegrí, and F. Rassoulzadegan. 1998. Consumption of picoplankton-size particles by marine ciliates: Effects of physiological state of the ciliate and particle quality. Limnol. Oceanogr. 43(3):458-464.

EPA (U.S. Environmental Protection Agency). 2008. Nanotechnology: Research Projects. National Center for Environmental Research. February 26, 2008. Available: http://es.epa.gov/ncer/nano/research/index.html [accessed October 16, 2008].

GAO (U.S. General Accountability Office). 2008. Report to Congressional Requesters Nanotechnology: Better Guidance is Needed to Ensure Accurate Reporting of Federal Research Focused on Environmental, Health, and Safety Risks. GAO-08-402. Washington, DC: U.S. General Accountability Office. March 2008 [online]. Available: http://www.gao.gov/new.items/d08402.pdf [accessed Aug. 26, 2008].

ICON (International Council on Nanotechnology). 2008. Towards Predicting Nano-Biointeractions: An International Assessment of Nanotechnology Environment, Health, and Safety Research Needs. International Council on Nanotechnology No. 4. May 1, 2008 [online]. Available: http://cohesion.rice.edu/CentersAndInst/ICON /emplibrary/ICON RNA_Report_Full2.pdf [accessed Aug. 26, 2008].

Jiang, J., G. Oberdörster, and P. Biswas. 2008a. Characterization of size, surface charge, and agglomeration state of nanoparticle dispersions for toxicological studies. J. Nanopart. Res. DOI 10.1007/s11051-008-9446-4.

Jiang, J., G. Oberdörster, A. Elder, R. Gelein, P. Mercer, and P. Biswas. 2008b. Does nanoparticle toxicity depend on size and crystal phase? Nanotoxicology 2(1):33-42.

Maynard, A. 2002. Experimental determination of ultrafine TiO_2 deagglomeration in a surrogate pulmonary surfactant: Preliminary results. Ann. Occup. Hyg. 46(Suppl. 1):197-202.

Maynard, A. 2008. Testimony to Committee on Science and Technology, U.S. House of Representatives: The National Nanotechnology Initiative Amendments Act of 2008, Annex A. Assessment of U.S. Government Nanotechnology Environmental Safety and Health Research for 2006. April 16, 2008 [online]. Available: http://democrats.science.house.gov/Media/File/Commdocs/hearings/2008/Full/16a pr/Maynard_Testimony.pdf [accessed Aug. 27, 2008].

Morfeld, P., and R.J. McCunney. 2007. Carbon black and lung cancer: Testing a new exposure metric in a German cohort. Am. J. Ind. Med. 50(8):565-567.

NEHI (Nanotechnology Environmental Health Implications Working Group). 2006. Environmental, Health, and Safety Research Needs for Engineered Nanoscale Materials. Arlington, VA: National Nanotechnology Coordination Office. September 2006 [online]. Available: http://www.nano.gov/NNI_EHS_research_needs.pdf [accessed Aug. 22, 2008].

NEHI (Nanotechnology Environmental Health Implications Working Group). 2007. Prioritization of Environmental, Health, and Safety Research Needs for Engineered Nanoscale Materials: An Interim Document for Public Comment. Arlington, VA: National Nanotechnology Coordination Office. August 2007 [online]. Available: http://www.nano.gov/Prioritization_EHS_Research_Needs_Engineered_Nanoscale_Materials.pdf [accessed Aug. 22, 2008].

NEHI (Nanotechnology Environmental Health Implications Working Group). 2008. National Nanotechnology Initiative Strategy for Nanotechnology-Related Environmental, Health, and Safety Research. Arlington, VA: National Nanotechnology Coordination Office. February 2008 [online]. Available: http://www.nano.gov/NNI_EHS_Research_Strategy.pdf [accessed Aug. 22, 2008].

NIOSH (National Institute for Occupational Safety and Health). 2005. FAQs About Control Banding. National Institute for Occupational Safety and Health. April 2005 [online]. Available: http://www.cdc.gov/niosh/topics/ctrlbanding/pdfs/CBFAQ.pdf [accessed July 30, 2008].

NRC (National Research Council). 1983. Risk Assessment in the Federal Government: Managing the Process. Washington, DC: National Academy Press.

NRC (National Research Council). 1996. Understanding Risk: Informing Decisions in Democratic Society. Washington, DC: National Academy Press.

Oberdörster, G., A. Maynard, K. Donaldson, V. Castranova, J. Fitzpatrick, K. Ausman, J. Carter, B. Karn, W. Kreyling, D. Lai, S. Olin, N. Monteiro-Riviere, D. Warheit, and H. Yang; ILSI Research Foundation/Risk Science Institute Nanomaterial Toxicity Screening Working Group. 2005. Principles for characterizing the potential human health effects from exposure to nanomaterials: Elements of a screening strategy. Part Fibre Toxicol. 2(1):8 doi:10.1186/1743-8977-2-8 [online]. Available: http://www.particleandfibretoxicology.com/content/2/1/8 [accessed Aug. 27, 2008].

Thomas, K. and P. Sayre. 2005. Research strategies for safety evaluation of nanomaterials, part I: Evaluating the human health implications of exposure to nanoscale materials. Toxicol. Sci. 87(2):316-321.

Thomas, K., P. Aguar, H. Kawasaki, J. Morris, J. Nakanishi, and N. Savage. 2006. Research strategies for safety evaluation of nanomaterials, part VIII: International efforts to develop risk-based safety evaluations for nanomaterials. Toxicol. Sci. 92(1):23-32.

5

Conclusions and Recommendations

The National Nanotechnology Initiative document *Strategy for Nanotechnology-Related Environmental, Health, and Safety Research* could be an effective tool for communicating the breadth of federally supported research associated with developing a more comprehensive understanding of the environmental, health, and safety implications of nanotechnology. It is the result of considerable collaboration and co-ordination among 18 federal agencies and is likely to eliminate unnecessary duplication of their research efforts.

***The Strategy for Nanotechnology-Related Environmental, Health, and Safety Research* does not describe a strategy for nano-risk research. It lacks input from a diverse stakeholder group, and it lacks essential elements, such as a vision and a clear set of objectives, a comprehensive assessment of the state of the science, a plan or road map that describes how research progress will be measured, and the estimated resources required to conduct such research.**

There remains an urgent need for the nation to build on the current research base related to the EHS implications of nanotechnology—including the federally supported research described in the 2008 NNI document—by developing a national strategic plan for nanotechnology-related environmental, health, and safety research.

Having reviewed the National Nanotechnology Initiative (NNI) strategy document, the committee has concluded that it does an excellent job of identifying numerous specific topics on which more research is needed to adequately address the environmental, health, and safety (EHS) concerns associated with engineered nanoscale materials. The committee found that, with some exceptions, the specific research needs in each research category were appropriate for nanotechnology-related EHS research. However, although the inventories of the research needs are sufficient for some research categories, they are poorly defined and incomplete in others, specifically risk management and exposure assessment. The committee also believes that some research needs that fall between categories could be overlooked.

The research needs in the NNI strategy document are not presented as concrete, measurable objectives, and the implementation plan fails to provide any sense of how success toward specific goals will be measured or what resources might be needed to achieve them.

The committee carefully considered the "gap analysis" in the NNI document, which was based on identifying FY 2006 funded projects as relevant to one or more of the five broad research categories. The committee concluded that the gap analysis is flawed and is neither accurate nor complete in laying a foundation for a research strategy. The approach used does not provide an accurate picture of current resource allocations even among the five broad categories. The committee concluded that the use of the FY 2006 data to conduct the gap analysis is perhaps the greatest flaw identified in the document. It is particularly problematic in the discussions of human health and metrology, in which it resulted in the inclusion of research projects that are not directly relevant to understanding the EHS needs related to nanomaterials. The issues arising from the gap analysis led to important deficiencies in all the research categories described in Section II of the 2008 NNI document. Because of the flaws in the gap analysis, it is difficult to understand the priorities of selected research needs and the logic for the priorities.

The NNI document states (p. 46) that "the EHS research strategy fundamentally depends on sustaining the broad spectrum of basic research. . . . The current balance of research funding addresses such basic investigations and supports regulatory decision making." However, although the committee has no reason to doubt the value of the compelling nanotechnology research described, it notes that probably less than half the grants and resources counted in the inventory will provide any useful data to support regulatory decision-making. The analysis suffers universally from a lack of coherent and consistent criteria for determining the value of information provided by various research activities. Such criteria would ideally be founded on an understanding of the uncertainties in each of the various research fields and the interrelationships among them.

The federal funding specifically addressing nanotechnology-related EHS issues is far less than portrayed in the NNI document and may be inadequate. The committee concludes that if no new resources are provided and the current agency funding continues, the implementation plan described in the NNI document will not ensure that engineered nanomaterials are adequately evaluated for potential health and environmental effects. Such an evaluation is critical to ensure that the future of nanotechnology is not burdened by uncertainties and innuendo about potential adverse health and environmental effects of engineered nanoscale materials. Those concerns have been voiced recently by both the nanotechnology industry and a variety of environmental and public-health interest groups.

In many endeavors, society looks to the scientific community for insights, data, and recommendations for establishing policies or regulations. In the broad swath of nanoscience and nanotechnology, the present committee considers that the emerging field of nanotechnology is one such endeavor. The scientific input needed for understanding the potential effects is not necessarily that produced by exploratory research (although it has its place) but rather often relies heavily on generating, identifying, and applying specific knowledge. In this respect, scientific input into developing policies for risk assessment and risk management of currently available and emerging nanotechnology bears a closer resemblance to the approval process for new drugs and medical devices than to the general advancement of new knowledge through exploratory research.

The current nanotechnology risk research portfolio is dominated by agencies traditionally focused on exploratory and investigator-driven research, such as the National Institutes of Health and the National Science Foundation. If these agencies are to continue to lead research efforts in this area, the scope of research requests and the review criteria used to assess the relative merits of submitted proposals may need to be modified if the agencies want to ensure that the research they support feeds into an effective EHS risk research strategy based on appropriate, targeted research.

There are several possible ways to accomplish such a change in criteria, for example, through joint initiatives, including requests for proposals with explicit statements of need, between federal agencies focused on fundamental or investigator-driven science and mission-driven agencies responsible for protecting human health and the environment (such as the Environmental Protection Agency, the National Institute for Occupational Safety and Health, the Food and Drug Administration, and the Consumer Product Safety Commission). Ultimately, any useful strategic plan for addressing EHS aspects of nanotechnology will have to focus on obtaining timely research results that can assist all stakeholders, including federal agencies, in planning, controlling, and optimizing the use of purposely engineered nanomaterials while minimizing and controlling the potential EHS effects of concern to society.

What is needed, the committee concludes, is an effective *national strategic plan* for nanotechnology-related EHS research that involves more stakeholders than the federal government. Such a plan would have to identify research needs clearly and estimate the financial and technical resources needed to address identified research gaps. A national strategic plan would be focused on providing solutions to challenges that do not necessarily fit neatly into disciplinary and institutional silos, and ensure important research does not fall between the gaps. Such a plan would also provide specific, measurable objectives and a timeline for meeting them.

The committee finds that the 2008 NNI document represents excellent input into a national strategic plan. A national strategic plan would ensure the timely development of engineered nanoscale materials that will bring about great improvements in the nation's health, its environmental quality, its economy, its security, and the quality of life without the unintended consequences of damage to the environment and to the health of the very workers and consumers who stand to benefit from the technology.

Reducing the burden of uncertainty through targeted, effective research that identifies and eliminates potential environmental and health hazards of engineered nanoscale materials should have high priority for the nation. An effective national strategic plan is essential for the successful development of and public acceptance of nanotechnology-enabled products. A value-of-information approach should be used to determine the research that is needed to reduce the current uncertainties with respect to the potential health and environmental effects of nanomaterials. A national strategic plan would need to address nanotechnology-based products that are entering commerce and nanotechnology-based products that are under development. It would provide a path for developing the scientific knowledge to support nanotechnology-related EHS risk-based decision-making. It would lay the scientific groundwork for addressing future materials and products arising out of new research, new tools, and new cross-fertilization between previously distinct fields of science and technology.

The committee chose the term *national strategic plan* rather than *federal strategic plan* because it concluded that one of the weaknesses of the 2008 NNI document is that it focuses only on federal government agency activities. Federal programs are essential and in the national interest, but the nongovernment research community should also contribute research and knowledge to the understanding of the EHS implications of nanotechnology.

The committee concludes that a truly national strategy cannot be developed within the limitations of the scope of research under the umbrella of the NNI. The NNI can produce only a strategy that is the sum of the individual agency priorities, many of which are not aligned with EHS research related to nanomaterials. The structure of the NNI makes the development of a visionary and authoritative research strategy extraordinarily difficult. Because the NNI is a coordination mechanism, not a research funding program, it has no central authority to make budgetary or funding decisions, and it relies on its member agencies to gather resources or influence to shape the overall federal nanotechnology-related EHS research activity. The NNI is responsible for ensuring U.S. competitiveness through the development of a rapid and robust nanotechnology-related research and development program while ensuring the safe and responsible development of nanotechnology itself, and these two missions may be perceived as being in conflict. But the conflict is a false dichotomy in that strategic research on potential risks posed by nanotechnology can be an integral and fundamental part of its sustainable development. Nonetheless, a clear separation of accountability for development of applications and assessment of potential EHS implications would help to ensure that the public-health mission is given appropriate priority.

Having considered those conclusions with respect to the 2008 NNI document and what is needed for a path forward, the committee offers the following two recommendations.

A robust national strategic plan is needed for nanotechnology-related environmental, health, and safety research that builds on the five categories of research needs identified in the 2008 NNI document. The development of the plan should include input from a broad set of stakeholders across the research community and other interested parties in government, nongovernment, and industrial groups. The strategy should focus on research to support risk assessment and management, should include value-of-information considerations, and should identify

- Specific research needs for the future in such topics as potential exposures to engineered nanomaterials, toxicity, toxicokinetics, environmental fate, and standardization of testing.
- The current state of knowledge in each specific area.
- The gap between the knowledge at hand and the knowledge needed.
- Research priorities for understanding life-cycle risks to humans and the environment.
- The estimated resources that would be needed to address the gap over a specified time frame.

As part of a broader strategic plan, NNI should continue to foster the successful interagency coordination effort that led to its 2008 document with the aim of ensuring that the federal plan is an integral part of the broader national strategic plan for investments in nanotechnology-related environmental, health, and safety research. In doing so, it will need a more robust gap analysis. The federal plan should identify milestones and mechanisms to ascertain progress and identify investment strategies for each agency. Such a federal plan could feed into a national strategic plan but would not itself be a broad, multistakeholder national strategic plan. Development of a national strategic plan should begin immediately and not await further refinement of the current federal strategy.

Appendix A

Biographic Information on the Committee for Review of the Federal Strategy to Address Environmental, Health, and Safety Research Needs for Engineered Nanoscale Materials

David L. Eaton *(Chair)* is associate vice provost for research at the University of Washington, where he holds faculty appointments as professor of environmental and occupational health sciences, professor of public health genetics, and adjunct professor of medicinal chemistry. He also serves as director of the University of Washington-National Institute of Environmental Health Sciences (NIEHS) Center for Ecogenetics and Environmental Health and directs a large, multi-investigator center grant from NIEHS in toxicogenomics. Dr. Eaton's research interests include the molecular basis of chemically induced cancers and the effect of human genetic variation in biotransformation enzymes on individual susceptibility to natural and synthetic chemicals. He was president of the Society of Toxicology in 2001-2002. He has served as chair of several National Research Council committees, including the Committee on Emerging Issues and Data on Environmental Contaminants and the Committee on EPA's Exposure and Human Health Reassessment of TCDD and Related Compounds. Dr. Eaton has been awarded many distinguished fellowships and honors, including the Achievement Award of the Society of Toxicology in 1990. He is an elected fellow of the Academy of Toxicological Sciences and of the American Association for the Advancement of Science. He earned his PhD in pharmacology and toxicology from the University of Kansas Medical Center.

Martin A. Philbert *(Vice Chair)* is a professor of toxicology and executive director of the Center for Risk Science and Communication at the University of Michigan. Dr. Philbert's research interests include the development of nanotechnology for intracellular measurement of biochemicals and ions and for the early detection of and treatment for brain tumors. He is actively engaged in the investigation of mechanisms of chemically induced energy deprivation syndromes in the central nervous system. He has published more than 100 scholarly manuscripts, book chapters, and abstracts and is the recipient of the 2001 Society of Toxicology Achievement Award. Dr. Philbert serves on the Institute of Medicine's Food and Nutrition Board and the Roundtable on Environmental Health Sciences, Research, and Medicine. He earned his PhD in neurochemistry and experimental neuropathology from the University of London.

George V. Alexeeff is deputy director for scientific affairs of the Office of Environmental Health Hazard Assessment (OEHHA) of the California Environmental Protection Agency. He oversees a staff of more than 80 scientists in multidisciplinary evaluations of the health effects of pollutants and toxicants in air, water, soil, and other media. His activities include reviewing epidemiologic and toxicologic data to identify hazards and derive risk-based assessments, developing guidelines to identify chemicals hazardous to the public, recommending air-quality standards, identifying toxic air contaminants, developing public-health goals for water contaminants, preparing evaluations for carcinogens and reproductive toxins, issuing sport-fish advisories, training health personnel on pesticide-poisoning recognition, reviewing hazardous-waste site risk assessments, and conducting multimedia risk assessments. He was chief of the Air Toxicology and Epidemiology Section of OEHHA from October 1990 through February 1998. Dr. Alexeeff has over 50 publications in toxicology and risk assessment. He has been a member of previous National Research Council committees including Evaluating the Efficiency of Research and Development Programs in the Environmental Protection Agency and the Scientific Review of the Proposed Risk Assessment Bulletin from the Office of Management and Budget. He earned his PhD in pharmacology and toxicology from the University of California, Davis.

Tina Bahadori is the managing director of the Long-Range Research Initiative (LRI) at the American Chemistry Council (ACC). Dr. Bahadori manages the development, implementation, and direction of the LRI research portfolios in environmental health with specific expertise and responsibilities in exposure and risk analysis. She is the LRI lead for developing a global research program on interpretation of biomonitoring data. Dr. Bahadori is the president-elect of the International Society of Exposure Analysis. She serves as an expert and reviewer on a number of scientific panels, including the National Academy of Sciences (NAS) panel for review of particulate-matter research; as a peer reviewer for the Environmental Protection Agency Science to Achieve Results grants; on the NAS interacademy panel on the ecology of the Caspian Sea; on the Advisory Panel for the Aerosol Research Inhalation Epidemiology Study; and on the internal steering committee and as one of the principal investigators for the St. Louis-Midwest PM Supersite. She was also a member of the Chemical Exposure Working Group for the National Children's Study. Before joining ACC, she was the manager for air-quality health integrated programs at the Electric Power Research Institute. Her research was related to health implications of environmental pollution and included integration of atmospheric chemistry, exposure assessment, and epidemiology. She was responsible for the design, implementation, and promotion of collaborative research with emphasis on policy and regulatory decision-making. At Arthur D. Little, Inc., where she was a consultant in the Environmental Risk Management Unit, she assisted clients with technical and management problems related to environment, health, and safety matters. She holds a doctorate in environmental science and engineering from the Harvard School of Public Health.

John M. Balbus is the chief health scientist and health program director at the Environmental Defense Fund and an adjunct associate professor of environmental health sciences at Johns Hopkins University. His expertise is in epidemiology, toxicology, and risk science. He spent 7 years at George Washington University, where he was the founding director of the Center for Risk Science and Public Health and served as acting chair of the Department of Environmental and Occupational Health. He was also an associate professor of medicine there. Dr. Balbus has served as a member of the Environmental Protection Agency (EPA) Children's Health Protection Advisory Committee, as a core peer consultation panel member for EPA's Voluntary Children's Chemical Exposure Program, and as a member of EPA review committees on air-toxics research, computational toxicology, and climate-change research. He serves on the National Research Council's (NRC) Board on Environmental Studies and Toxicology and is a member of the Committee on Improving Risk Analysis Approaches Used by the U.S. EPA. He previously served on the NRC Committee on Applications of Toxicogenomics Technologies to Predictive Toxicology. Dr. Balbus received his MD from the University of Pennsylvania.

Moungi G. Bawendi (NAS) is the Lester Wolfe Professor of Chemistry in the Department of Chemistry at the Massachusetts Institute of Technology. His research interests include the chemistry, physics, and applications of nanometer-size semiconductor and metal particles exhibiting quantum mechanical size effects. He is interested in the science and applications of nanocrystals, especially semiconductor nanocrystals. Previously, he was a member of the National Research Council Committee on the Review of the National Nanotechnology Initiative. In 2007, Dr. Bawendi was elected to the National Academy of Sciences. He earned a PhD in chemistry from the University of Chicago.

Pratim Biswas is the Stifel and Quinette Jens Professor and chair of the Department of Energy, Environmental and Chemical Engineering at Washington

University in St. Louis. Dr. Biswas's research interests include aerosol science and engineering, nanoparticle technology, air quality and pollution control, combustion, environmentally benign energy production and materials processing (with applications in environmental and energy technologies), and thermal sciences (heat transfer and fluid mechanics). Dr. Biswas was appointed president of the American Association for Aerosol Research for 2006-2007 and had been the technical program chair at the International Aerosol Conference in St. Paul, MN. He received his PhD in mechanical engineering from the California Institute of Technology.

Vicki Colvin is professor of chemistry at Rice University and director of its Center for Biological and Environmental Nanotechnology (CBEN). CBEN is one of the nation's six nanoscience and engineering centers funded by the National Science Foundation. One of CBEN's primary interests is the application of nanotechnology to the environment. Dr. Colvin has received numerous accolades for her teaching abilities, including Phi Beta Kappa's Teaching Prize for 1998-1999 and the Camille Dreyfus Teacher Scholar Award in 2002. In 2002, she was also named one of Discover magazine's "Top 20 Scientists to Watch" and received an Alfred P. Sloan Fellowship. Dr. Colvin is a frequent contributor to Advanced Materials, Physical Review Letters, and other peer-reviewed journals and holds patents to four inventions. She received her PhD in chemistry from the University of California, Berkeley, where she was awarded the American Chemical Society's Victor K. LaMer Award for her work in colloid and surface chemistry.

Stephen J. Klaine is a professor in the Department of Biological Sciences and the director of the Institute of Environmental Toxicology at Clemson University. His research focuses on the fate and effects of contaminants in the environment, specifically contaminants that migrate from various land uses into aquatic ecosystems and their effects on aquatic plants and animals. His laboratory studies contaminant effects on fish, aquatic invertebrates, plants, and algae. Currently, it is studying the toxicity of metals, pesticides, pharmaceuticals, and nanomaterials. Dr. Klaine received the Sigma Xi Researcher of the Year Award at Clemson University in 2007 and has been named to Who's Who in Technology, Environmental Science and Engineering. He has served on the National Research Council Panel on Life Sciences and is aquatic-toxicology editor for Environmental Toxicology and Chemistry. He received his PhD in environmental science from Rice University.

Andrew D. Maynard is the chief science adviser at the Woodrow Wilson International Center for Scholars for the Project on Emerging Nanotechnologies. He also holds an associate professorship at the University of Cincinnati and is an honorary senior lecturer at the University of Aberdeen, UK. Dr. Maynard's research interests revolve around aerosol characterization and the implications of nanotechnology for occupational health. His expertise covers many facets of aerosols and health implications, from occupational aerosol sampler design to state-of-the-art nanoparticle analysis. Previously, he worked for the National Institute for Occupational Safety and Health (NIOSH) and represented the agency on the Nanoscale Science, Engineering and Technology Subcommittee (NSET) of the National Science and Technology Council; he also cochaired the Nanotechnology Health and Environment Implications Working Group of the NSET. Recently, he was a recipient of the NIOSH Alice Hamilton Award (Biological Sciences). He is a member of the Executive Committee of the International Council on Nanotechnology and until recently chaired the International Standards Organization working group on size-selective sampling in the workplace. He earned his PhD in aerosol physics from the Cavendish Laboratory, University of Cambridge, UK.

Nancy Ann Monteiro-Riviere is a professor of investigative dermatology and toxicology at the Center for Chemical Toxicology Research and Pharmacokinetics at North Carolina State University (NCSU). Dr. Monteiro-Riviere is also a professor in the Joint Department of Biomedical Engineering of the University of North Carolina (UNC)-Chapel Hill and NCSU and research adjunct professor of dermatology in the School of Medicine at UNC-Chapel Hill. She is a past president of the Dermal Toxicology Specialty Section and the In Vitro Toxicology Specialty Section of the Society of Toxicology. Dr. Monteiro-Riviere is a fellow of the American Academy of Nanomedicine, the Academy of Toxicological Sciences, and the American College of Toxicology. She serves on several toxicology editorial boards and national panels, including many in nanotoxicology, and she is coeditor of *Nanotoxicology: Characterization and Dosing and Health Effects*. She received her PhD in anatomy from Purdue University.

Günter Oberdörster is professor in the Department of Environmental Medicine at the University of Rochester, director of the University of Rochester Ultrafine Particle Center, principal investigator on a multidisciplinary research initiative in nanotoxicology, and head of the Pulmonary Core of a National Institute of Environmental Health Sciences center grant. His research focuses on the effects and underlying mechanisms of lung injury induced by inhaled nonfibrous and fibrous particles, including extrapolation modeling and risk assessment. His studies of ultrafine particles influenced the field of inhalation toxicology, raising awareness of the unique biokinetics and toxic potential of nanoscale particles. He has served on many national and international committees and is a recipient of several scientific awards. He is on the editorial boards of the Journal of Aerosol Medicine, Particle and Fibre Toxicology, Nanotoxicology, and the International Journal of Hygiene and Environmental Health and is associate editor of *Inhalation Toxicology and Environmental Health Perspectives*. He earned his DVM and PhD (in pharmacology) from the University of Giessen, Germany.

Mark A. Ratner (NAS) is the Morrison Professor of Chemistry and professor of materials science and engineering at Northwestern University. His research focuses on structure and function at the nanoscale and on the theory of fundamental chemical processes. Specific interests include molecular electronics, electron transfer, self-assembly, nonlinear optical response in molecules, and theories of quantum dynamics. He is a fellow of the American Association for the Advancement of Science and a member of the American Academy of Arts and Sciences. Dr. Ratner was elected to the National Academy of Sciences in 2002 for his contributions to molecular materials theory and modeling. He earned his PhD in chemistry from Northwestern University.

Justin G. Teeguarden is a senior research scientist with the Pacific Northwest National Laboratory where he conducts research within a multidisciplinary team studying the relationship between the physicochemical properties of nanomaterials and their biocompatibility. His major research focus is in the areas of nanomaterial pharmacokinetics and dosimetry, both *in vivo* and *in vitro*, and the development of integrated computational models of cellular and tissue dosimetry and biologic response. He is the principal investigator of pharmacokinetic studies of organic chemicals and metals and develops physiologically based pharmacokinetic models of chemical kinetics for application in study design and risk assessment for both private companies and the EPA. Through Society of Toxicology symposia, specialty sections and continuing education courses, Dr. Teeguarden has promoted the application of the fundamental sciences in nanomaterial risk assessment. He serves on the National Toxicology Program Board of Scientific Councilors, and on a variety of EPA and NIH review panels. Dr. Teeguarden received his PhD in toxicology from the University of Wisconsin, Madison, and is board certified in toxicology.

Mark R. Wiesner is the James L. Meriam Professor of Civil and Environmental Engineering in the Pratt School of Engineering at Duke University. He was previously Chair of Excellence in the Chemical Engineering Laboratory at the Institute Nationale Polytechnique, Toulouse, France. His research interests include membrane processes, nanostructured materials, transport and fate of nanomaterials in the environment, colloidal and interfacial processes, and environmental systems analysis. Dr. Wiesner has received the Association of Environmental Engineering and Science Professors Frontiers in Research Award, the American Institute of Chemical Engineers Graduate Research Award for Membrane-Based Separations, and the Charles Duncan Award for Scholarship and Teaching at Rice University. He served on the Scientific Advisory Board and was the U.S. director for the European Union-United States University Consortium on Environmental Engineering Education from 1993 to 2005. Dr. Wiesner received his PhD in environmental engineering from Johns Hopkins University.

Appendix B

Statement of Task

The National Research Council shall conduct a scientific and technical review of the draft document entitled "Federal Strategy for Environmental, Health, and Safety (EHS) Research Needs for Engineered Nanoscale Materials," expected to be publicly released by the U.S. National Nanotechnology Initiative in September 2007. An ad hoc committee will plan a workshop and evaluate the scientific and technical aspects of the draft strategy and comment in general terms on how this strategy will develop information needed to support the EHS risk assessment and risk management needs with respect to nanomaterials. In its evaluation the committee will take into consideration the report, *Environmental, Health, and Safety Research Needs for Engineered Nanoscale Materials* (NEHI 2006) and other governmental and non-governmental reviews identifying EHS research priorities.

Appendix D

National Nanotechnology Initiative Strategy for Nanotechnology-Related Environmental, Health, and Safety Research[1]

The National Nanotechnology Initiative's *Strategy for Nanotechnology-Related Environmental, Health, and Safety Research* is available on CD-ROM on the inside of the back cover of this report.

---

[1]NEHI (Nanotechnology Environmental Health Implications Working National Nanotechnology Initiative Strategy for Nanotechnology-Rela Health, and Safety Research. Arlington, VA: National Nanotech Office. February 2008 [online]. Available: http://www.nano.g Strategy.pdf [accessed Aug. 22, 2008].

Appendix C

Workshop Agendas of the National Research Council Committee for Review of the Federal Strategy to Address Environmental, Health, and Safety Research Needs for Engineered Nanoscale Materials

1st Workshop: March 31, 2008

Lecture Room
National Academy of Sciences
2100 C Street, NW
Washington, DC

Public Agenda

2:30 PM   Welcome and Introductory Remarks
David Eaton, Chair, NRC Committee for Review of the Federal Strategy to Address Environmental, Health, and Safety Research Needs for Engineered Nanoscale Materials

2:40 PM   Dr. Clayton Teague – Introduction to the NNI Strategy and Expectations for the Review
Director, National Nanotechnology Coordination Office

Committee – Panel Discussion with Members of the Nanotechnology Environmental Health Implications (NEHI) Working Group

Environmental Protection Agency
Jeffrey Morris, *Acting Director, Office of Science Policy*
Phillip Sayre, *Associate Director, Risk Assessment Division, Office of Pollution, Prevention, and Toxics*

Food and Drug Administration
Norris Alderson, *Associate Commissioner for Science*
Richard Canady, *Senior Science Policy Analyst, Office of the Commissioner*

National Institute of Environmental Health Sciences
Sally Tinkle, *Senior Science Advisor*

National Institute for Occupational Safety and Health
Paul Schulte, *Director, Education and Information Division*
Vladimir Murashov, *Special Assistant to the Director*

National Institute of Standards and Technology
Dianne Poster, *Policy Analyst*

4:40 PM	Public Comments
5:30 PM	Adjourn Public Session

2nd Workshop: May 5, 2008

Room 100
Keck Center of the National Academies
500 5th Street, NW
Washington, DC

Public Agenda

8:45 AM	Welcome and Introductory Remarks *David Eaton, Chair, NRC Committee for Review of the Federal Strategy to Address Environmental, Health, and Safety Research Needs for Engineered Nanoscale Materials*
9:00 AM	Discussion of the development of the EU framework for EHS research on nanotechnology (videoconference)

Dr. Philippe Martin, *Directorate General, Health and Consumer Protection, European Commission*
Dr. Pilar Aguar, *Program Officer, Directorate General for Research, European Commission*

10:00 AM	Committee – Panel Discussion on the Federal Strategy to Address Environmental, Health, and Safety Research Needs for Engineered Nanoscale Materials Carolyn Cairns, *Program Leader, Product Safety, Consumer's Union* Thomas Epprecht, *Director of Products, Swiss Re* (video conference) William Gulledge, *Senior Director, Chemical Products and Technology Division, ACC* Michael Holman, *Research Director, Lux Research* William Kojola, *Industrial Hygienist, AFL-CIO* Terry Medley, *Global Director of Corporate Regulatory Affairs, DuPont* Jennifer Sass, *Senior Scientist, Natural Resources Defense Council*
1:15 PM	Committee – Panel Discussion on the influence of the Federal strategy on decision making and priority setting Altaf Carim, *Program Manager, Office of Science, Department of Energy* William Rees, *Deputy Under Secretary of Defense for Laboratories and Basic Sciences* Mihail Roco, *Senior Advisor for Nanotechnology, National Science Foundation*
2:15 PM	Public Comments
3:00 PM	Adjourn Public Session

Appendix C

Workshop Agendas of the National Research Council Committee for Review of the Federal Strategy to Address Environmental, Health, and Safety Research Needs for Engineered Nanoscale Materials

1st Workshop: March 31, 2008

Lecture Room
National Academy of Sciences
2100 C Street, NW
Washington, DC

Public Agenda

2:30 PM	Welcome and Introductory Remarks *David Eaton, Chair, NRC Committee for Review of the Federal Strategy to Address Environmental, Health, and Safety Research Needs for Engineered Nanoscale Materials*
2:40 PM	Dr. Clayton Teague – Introduction to the NNI Strategy and Expectations for the Review *Director, National Nanotechnology Coordination Office*

Committee – Panel Discussion with Members of the
Nanotechnology Environmental Health Implications (NEHI)
Working Group

Environmental Protection Agency
Jeffrey Morris, *Acting Director, Office of Science Policy*
Phillip Sayre, *Associate Director, Risk Assessment Division,
Office of Pollution, Prevention, and Toxics*

Food and Drug Administration
Norris Alderson, *Associate Commissioner for Science*
Richard Canady, *Senior Science Policy Analyst, Office of the
Commissioner*

National Institute of Environmental Health Sciences
Sally Tinkle, *Senior Science Advisor*

National Institute for Occupational Safety and Health
Paul Schulte, *Director, Education and Information Division*
Vladimir Murashov, *Special Assistant to the Director*

National Institute of Standards and Technology
Dianne Poster, *Policy Analyst*

4:40 PM   Public Comments

5:30 PM   Adjourn Public Session

2nd Workshop: May 5, 2008

Room 100
Keck Center of the National Academies
500 5th Street, NW
Washington, DC

Public Agenda

8:45 AM   Welcome and Introductory Remarks
               *David Eaton, Chair, NRC Committee for Review of the
               Federal Strategy to Address Environmental, Health, and
               Safety Research Needs for Engineered Nanoscale Materials*

9:00 AM   Discussion of the development of the EU framework for EHS
               research on nanotechnology (videoconference)

Foodie
CITY BREAKS
>*EUROPE*

25 CITIES, 250 ESSENTIAL EATING EXPERIENCES

RICHARD MELLOR

DOG 'n' BONE

About the author

A self-described "gun for hire," Richard Mellor is
a freelance travel and food journalist who writes
regularly for *Metro*, *The Times*, *The Guardian*,
Telegraph Online, the *Evening Standard*, and
Broadly. He is based in London, UK.

Published in 2018 by Dog 'n' Bone Books
An imprint of Ryland Peters & Small Ltd
20–21 Jockey's Fields, London
WC1R 4BW

341 E 116th St, New York, NY 10029

www.rylandpeters.com

10 9 8 7 6 5 4 3 2 1

Text © Richard Mellor 2018
Design © Dog 'n' Bone Books 2018

A CIP catalog record for this book is
available from the Library of Congress
and the British Library.

ISBN: 978 1 911026 48 8

Printed in China

Editor: Caroline West
For picture credits, see page 144

Contents

Introduction

> Once upon a time, city breaks revolved around culture—Paris for the Eiffel Tower, Barcelona to climb Gaudi's Sagrada Família, etc.—and food was something fitted in around that; a quirky, limited-research experience of mysterious ingredients and contentious cooking.

How times have changed. By 2016, a Food Travel Monitor report had 75 percent of leisure travelers admitting that they chose their destination for culinary reasons. It's just one of numerous surveys from the past decade which all roughly prove the same, now-universal truth: that gastronomy has surpassed all other tourist requirements. Today, rather than merely being a consideration around why and where we travel, food and drink experiences are, often, the crucial factor in our vacation-decision-making.

Why is that? The rise of low-cost travel, led by no-frills airlines, is one key element, making quickie city breaks increasingly tempting and conceivable. Then there's the sharing-economy likes of Airbnb, VizEat, Vayable, and Spinlister: not only do these platforms conspire to reduce trip costs, but they encourage people to meet and live like locals—to really deep-dive into a destination, steep oneself in it.

As well as the Eiffel Tower and Sagrada Família, modern-day travelers wander through organic markets, take classes with artisan mask-makers, go jogging in the company of fitness bloggers, join street-art tours, or clink cocktails with the neighborhood dominatrix. Immersion is all, and that extends to food. Not only do we want to eat the best food, but we want to dine like resident Berliners or Neapolitans and sample products typical to their environment. We want to taste the destination.

And that, roughly, is the thinking behind this book—which concentrates on the cities offering Europe's premier epicurean thrills. While including luxurious, budget-blowing suggestions—there will always be a place for these—I've also proffered cheapo (but brilliant) alternatives, oddball brasseries in atypical districts, third-wave coffee bars, ice-cream legends, revived markets, and specialist tavernas. In other words, venues which, when experienced together, begin to offer a representative taste of each city.

Finally, a brief explanation of how this book works. Each of the 25 chosen cities is subdivided into ten recurring sections. These include "Caffeine Kicks" (i.e. cafés), "Market Research" (food markets), "Local Secret" (restaurants unknown to tourists), and "Hip & Happening" (the trendy place to go).

I hope you enjoy them as much as me, and that the book proves repeatedly useful. Happy eating!

Richard Mellor

The Cities

Copenhagen

DENMARK

> Almost a decade ago, Copenhagen hit the foodie big time: Noma was the world's best restaurant, and New Nordic its most exalted cuisine. Though that buzz has inevitably diminished since, Denmark's capital remains a gastro-must. Few cities can match its cocktails or coffee, while the cooking remains as thrilling as ever.

ON A BUDGET?

Everyone will tell you: Copenhagen is expensive. And they're right, what with the 50-kroner cappuccinos and 160-kroner burgers. But there are ways to keep costs down, like eating the vegetarian soup of the day at *Café Ørstrup* in Indre By. It costs 52 kroner with husks of organic bread; cheaper still is a pork stew or the open sandwiches on rye bread. Fully vegetarian is the longstanding *Morgenstedet* restaurant, which is located in a pretty cottage in the hippie "freetown" of Christiania. Navigate the marijuana fumes and eat coconut curries for around 100 kroner.

Holbergsgade 22 (+45 28 55 1888, www.cafeorstrup.dk); Fabriksområdet 134 (no phone, no website)

SPLASH OUT

While *Noma*'s new farm-like venue in Christiania ensures the restaurant continues to dominate the headlines, there are plenty of other fine-dining establishments worthy of your attention in the city. *Relæ*, *Amass*, the Thai-focused *Kiin Kiin*, *Geranium*, and new *108* are foremost. But, for something altogether more exciting, board the 20-minute flight (you're splashing out after all) to Bornholm. Day-trippers to this sedate, beautiful island can eat at the original, beachside **Kadeau** (a second sister venue is open back in Copenhagen). Using the isle's best ingredients, dishes such as oysters with parsley, sauerkraut, and blackcurrant leaves are interesting, wholesome, and vibrant, but never pretentious.

Baunevej 18, Vestre Sømark, Bornholm (+45 56 97 8250, www.kadeau.dk/bornholm.php)

HIP & HAPPENING
Nørrebro is Copenhagen's coolest quarter. Most recently gentrified and still somewhat raw is its southerly Meatpacking District: once a haunt of sex-workers and troublemakers, but now the voguish preserve of cool wine bars and street-food stalls. The best place is broad **Kødbyens Fiskebar**, where ex-Noma sommelier Anders Selmer serves up superb oysters, Limfjorden mussels, or even, yes, fish and chips! Mains are unexpectedly affordable, too. Fancier and pricier is **Scarpetta**, some way north. The traditional Italian fare there

is accompanied by Hans Wegner's cool Y chairs.

Flæsketorvet 100 (+45 32 15 5656, www.fiskebaren.dk/ fiskebaren); Rantzausgade 7 (+45 35 35 0808, www.cofoco.dk/#restaurants)

▼ Amass is just one of several top restaurants run by former chefs at Noma. This means provenance and seasonality have come to play a key role in shaping the city's outlook on food.

LOCAL SECRET

Anywhere championed by Noma honcho René Redzepi gets our vote. Despite those regular plugs, however, *Café det Vide Hus* continues to operate largely under the tourist radar. Perhaps its limited food tariff is to blame. The café, which has lovely views of Rosenborg Castle, only offers snack-type stuff: avocado on rye toast, cookies, scones, and so forth. Best are its wacky homemade desserts, such as sea buckthorn sorbet dipped in white chocolate or elderflower ice cream with chocolate and bee pollen. Up a spiral staircase you'll find a snug lounge for quality espressos.

Gothersgade 113
(+45 60 61 2002,
www.facebook.com/detvidehus)

REGIONAL COOKING

Denmark's classic snacks are smørrebrød—one-sided open sandwiches made with chewy rye-bread. City-center *Palaegade* serves over 40 varieties at lunchtimes, from the truly vintage herring to tenderloin with creamy mushrooms. Those who prefer to seek out the classics book for lunch at *Sankt Annæ* (described by famous British food critic AA Gill as almost faultless), but to taste the latest

▶ Smørrebrød is an ever-present across Denmark, but you'll find the most variety in the capital—from traditional at Sankt Annæ (a favorite of the Danish Royal Family) to modern at Vækst.

New Nordic fads, try the aforementioned *108* or eye-catching *Vækst*. Thrillingly set in a leafy glasshouse to "capture the nature of the garden party and prolonged feel of Danish summer," *Vækst's* pair of rooms also star upcycled furniture. Food, much of it grown on site, stretches to the likes of octopus beside cheese, chicken skin, and kohlrabi, and geranium ice cream with porridge.

Palægade 8 (+45 70 82 8288,
www.palaegade.dk);
Sankt Annæ Plads 12
(+45 33 12 5497
www. restaurantsanktannae.dk
Sankt Peders Stræde 34
(+45 38 41 2727,
 www.cofoco.dk/vaekst)

▲ The compact size of Copenhagen means a stroll through its pretty streets will soon reveal an enticing place to eat.

Jagtvej 147 (+45 35 83 7213, www.tagensborg.dk); Ægirsgade 10 (+45 35 83 2916, no website)

🍴 BRILL FOR BRUNCH

Head to *Toldboden*, a single-storey, isolated building set scenically in Copenhagen's former ferry terminal, for the chance to have Danish-like breakfasts. Brunch menus at this popular harbor space change monthly for reasons of seasonality, but you'll always find rye breads, Danish cheeses, smoked salmon, shrimps, and hot waffles. Yogurts with homemade granola also feature, as do bacon and eggs. They even make their own chocolate spread, and everything can be washed down with good coffee and cold-pressed apple juices. Deckchairs right by the water offer an option for post-meal slumber. Book ahead.

Nordre Toldbod 18–24 (+45 33 93 0760, www.toldboden.com)

📋 QUINTESSENTIALLY COPENHAGEN

Musty, murky, smoky, poky. Plentiful across the city, Copenhagen's bodega pubs are like nowhere else. They range from slow-paced, sparsely populated, suburban examples where locals play cards to more central ones, whose cheap drinks and late-night hours lure students. One of the latter is *Tagensborg*, which sells bargain bottles of Tuborg amid its tobacco fog. Also providing a form of hygge is tiny *Karusellen* in Nørrebro. There, Faroese sailors chase beers with glasses of Dr. Nielsen a fearsome local schnapps, as the jukebox jolts out soul ballads.

🍸 KILLER COCKTAILS

Yes, you'll blow your budget, but there's simply no missing *Ruby* for cocktail disciples. Amid 1920s decor and leathery, lounge-style chairs, expert bartenders prepare a huge variety of drinks from classic to homemade. One of the new creations is The Dark Side, featuring rich Edition Negra rum, elderberry liqueur, and smoky Laphroaig whisky, all served on ice. Another is Real Sweet Pear The 2nd, where Tanqueray gin meets local pear liqueur, crème de cacao, and a dollop of fresh organic cream. Table service is offered, and the warmth is heavenly of a chilly night.

Nybrogade 10 (+45 33 93 1203, www.rby.dk)

☕ CAFFEINE KICKS

There are cool, craft java destinations across Copenhagen, but foremost is *The Coffee Collective*'s outpost on Jægersborggade. After roasting premium, directly sourced coffee beans for numerous cafés, the owners opened their own place in Nørrebro. It's relaxed and handsome, the roaster right there in front of customers. Just down the same street, off handsome Assistens Kirkegård park and lined with restaurants and boutiques, is *Meyer's Bageri*. Its takeaway *snegle*— cinnamon-spiked, chocolate-coated, snail-shaped pastries —go perfectly with coffee.

Jægersborggade 57 (+45 60 15 1525, www.coffeecollective.dk)

🏬 MARKET RESEARCH

There's another newer Coffee Collective branch at *Torvehallerne* (Mon–Fri, 10am–6pm or later; Sat and Sun, 11am–5pm), two neighboring, glass-encased halls in which 60 stands hawk fresh fish and meat, gourmet chocolate, vegetables, classic licorice, and all sorts of exotic spices. Plenty of seating areas and bike stands make this a real social hub, as do lots of restaurant stalls. Just beware the emboldened crows. Standouts? Try Hija de Sanchez's crispy cod tacos, Gorm's artisan pizza, and cozy Tapa del Toro's meatballs under Manchego cheese. Beer lovers should visit Mikkeller & Friends Bottleshop. Here you can stock up on ales produced by the eponymous Mikkeller, Denmark's top craft brewer.

Frederiksborggade 21 (no phone, www.torvehallernekbh.dk)

◀ Torvehallerne has become a one-stop shop for foodies.

Stockholm

SWEDEN

> While Copenhagen was the early HQ of New Nordic, Stockholm soon caught up. Today the city, which comes alive in summer, has Michelin stars to spare. Yet it also has hot-dog stands, grand market halls, great coffee, and those famous cinnamon buns. You'll eat well there.

ON A BUDGET?

Although Sweden isn't as pricey as other Scandinavian countries, it's hardly a bargain-priced destination for a weekend away. To save money, look to two distinct culinary boom areas. First comes Asian food: huge there, and packed with bargain lunch buffets where the all-inclusive pricing means you can refill, refill, and refill some more. Try ***Bamboo Palace***, where there are 95-kronor offers separated into general Japanese, sushi, and Chinese options. Alternatively, consider Stockholm's beloved korvar (hot dogs), which come stuffed into piles of mash and roasted onions. ***Korvkiosk***'s branch at the Teatern market is dirt cheap.

Kungsgatan 17 (+46 8 21 8241, www.bamboopalace.se); Götgatan 100 (no phone, www.korvkiosk.se)

SPLASH OUT

Cocktail and food mash-ups are de rigueur in Stockholm, but Östermalm venue ***Penny & Bill*** goes farther by offering direct pairings. You don't have to order their combinations, but a taste extravaganza is assured by doing so. That could be pork belly and soy-cured yolk alongside an On The Decks (dark rum and ginger beneath a Scotch, vanilla, lime, and soda float). Or sip a Juicy Goose, featuring white rum, cloves, and pineapple, between

▲ Dishes on offer at Penny & Bill, which describes itself as "a symbiosis of restaurant and cocktail bar."

mouthfuls of zander ceviche and cilantro- (coriander-) oyster dressing. Desserts incorporate local produce: cloudberry sorbets with dashes of absinthe and rye-bread apple crumbles.

Grev Turegatan 30
(+46 8 611 0211,
www.pennyandbill.se)

HIP & HAPPENING

The original Punk Royale restaurant is famed not only for its "bizarro" cuisine and drinks served from old jerry cans, but also for being tremendously tough to book. Accordingly, its owners have opened the *Punk Royale Café* next door. On offer are smaller, cheaper "snacks" borrowed from the parent venue's tasting menus, with names like "Craving" and "Horny." You can gobble foie gras on buttery toast (under ketchup smileys, indeed), or potato-and-almond pancakes with fish roe and fresh cream. A smoke machine pipes out fog and pork chops hang from the shabby-chic roof.

Folkungagatan 128 (+46 8 128 22411, www.punkroyale.se)

LOCAL SECRET

Stockholm boasts a serious number of French restaurants. Surely none are quite so Gallic as *Bistrot Paname*, where old Parisian street signs, car plates, soccer scarves, and the odd Tricolour cover the walls. The menu is equally patriotic, right down to its frogs' legs and garlicky escargot. Boeuf bourguignon is the most popular order, just ahead of sausage and lentils. Also present and correct are omelets, imported cheeses, choucroute, pastis, champagnes, and chardonnays. *Très bien, très bien*. The vibe is slightly boho and very *sympathique*.

Hagagatan 5 (+46 8 31 4338, no website)

BRILL FOR BRUNCH

Fancy brunch on a boat? That's what's on offer with *S/S Stockholm*, whose three-hour brunch cruises see passengers dining as the vessel glides around on weekends. The spread is a sort of international buffet, ranging from local herring to croissants and homemade quiche. You can stay in the grand dining room throughout, or slumber in lounges. Back on land, in well-to-do Östermalm, *Nybrogatan 38* has scrambled eggs, croque madams, chia puddings, and lots of yogurts to investigate. Cinnamon buns, a Swedish classic, come fried and supported by gooseberry or strawberry jelly (jam) and cream.

Kajplats 16 (+46 8 1200 4000, www.stromma.se/stockholm); Nybrogatan 38 (+46 8 662 3322, www.nybrogatan38.com)

🍲 REGIONAL COOKING

Another staple is meatballs, or *köttbulle*. They're carefully handmade from a variety of impeccably sourced meats and fish at deli-restaurant ***Meatballs for the People***. Most exotic are moose and fennel balls, supported by mashed potato, cream sauce, lingonberry jus, and pickled cucumber. But if you want to go all out Swedish, make for ***Pelikan***, in trendy Södermalm. Its proud, old-school plates include *fläsklägg med rotmos*: cured ham hock, mashed carrots, turnips, and potatoes, and a spoonful of sweet mustard. You'll sit under beautiful wood-paneling in a room as uncomplicated and sincere as the cuisine.

Nytorgsgatan 30 (+46 8 466 6099, www.meatball.se); Blekingegatan 40 (+46 8 556 09090, www.pelikan.se)

✅ QUINTESSENTIALLY STOCKHOLM

New Nordic remains firmly in vogue in Stockholm, but so do stresses on healthy and locavore cooking. Hence Östermalm's new ***Gastrologik***, whose well-trained chef-owners Jacob and Anton beautifully execute a long, seasonal, and hugely inventive tasting menu.

▲ Meat and potatoes, Swedish style.

The restaurant closes on Mondays, specifically so they can rummage around a biodynamic vegetable garden up in Rosendal. Named Sweden's "Chef of the Year" in 2014, Filip Fastén maintains a similar philosophy at the small, less formal ***Agrikultur***, in the northern suburbs. Many ingredients have been foraged, fermented, brewed, planted, or handpicked by Filip and co-chef Joel.

Artillerigatan 14 (+46 8 662 3060, www.gastrologik.se); Roslagsgatan 43 (+46 8 15 0202, www.agrikultur.se)

🍸 KILLER COCKTAILS

There are two main reasons to drink at the ***Reisen Bar***. First, its scenic location: on the waterfront in old Gamla Stan. Second, its super staff: experienced mixologists, which means outstanding cocktails. Part of an eponymous hotel, Reisen's offerings include From Yokohama With Love, starring Japanese plum liqueur, ginger and soy rock candy, grapefruit bitters, and aquavit, a Danish schnapps made using caraway seeds. More of a sneaker-wearer? Try ***Trädgården***. Along with burger stands and a bocce court, this all-night summer party space in far southern Södermalm incorporates a chill cocktail bar.

Skeppsbron 12 (+46 8 545 13991, www.firsthotels.com); Hammarby Slussväg 2 (+46 8 644 2023, www.tradgarden.com)

☕ CAFFEINE KICKS

Cool coffee shops are ten a krona here, but two stand out—both in Södermalm. ***Drop Coffee***'s managing director Joanna Alm is a former runner-up at the "World Coffee Roasting Championship," which guarantees the homely bar knows what its doing. Besides the sustainable beans and floral brews, this is a prime place in which to practice *fika*—that Swedish concept of relaxing coffee breaks. Likewise, the high-ceilinged **Johan & Nyström**, whose teas and coffees have been sourced from around the globe. Coffee accessories for reciprocal home-brewing are on sale, and there's a training lab below.

Wollmar Yxkullsgatan 10 (+46 8 410 23363, www.dropcoffee.com); Swedenborgsgatan 7 (+46 8 702 2040, www.johanochnystrom.se/ konceptbutiken)

MARKET RESEARCH

Which food hall to frequent? There are plenty to choose from. The most acclaimed one is **Östermalmshallen** (Mon–Fri, 9.30am–7pm; Sat, 9.30am–5pm), a red-brick building that dates back to the 1880s. Seafood and local fodder, summarily called *husmanskost*, are the main focuses, while Jamie Oliver's a famous fan. At the time of writing, the hall is being renovated until spring 2019 so, until then, visit Norrmalm's **Hötorgshallen** (Mon–Thurs,

10am–6pm; Fri, 10am–7pm; Sat, 10am–4pm) instead. You'll see ox tongues, crayfish fillets, Västerbotten cheeses, and glistening chanterelles all being loudly bargained for.

Östermalmstorg (no phone, www.ostermalmshallen.se); Sergelgatan 29 (+46 8 23 00 01, www.hotorgshallen.se)

▼ Östermalmshallen has been serving quality food to appreciative Stockholm residents for over 125 years.

Edinburgh

SCOTLAND

> Whether you're inspecting the Scottish Parliament, trudging up Arthur's Seat, attending the Fringe, or admiring the Castle, Edinburgh can be an exhausting place. You'll need fuel, and no mistake. Luckily, whether it's haggis, hog-roast rolls, or hake with a miso glaze, the city has you covered.

💰 ON A BUDGET?

There are three reasons why **Bells Diner**, secreted in handsome Stockbridge, excels. Firstly, it has been going for 40 years, so clearly has this hospitality thing cracked. Next, the burgers (there are, supposedly, steaks and salads on the menu, but ordering them would be akin to requesting pizza in KFC) are superb and superbly filling, no matter which size and topping combination you gleefully opt for. Even more so in cahoots with a yummy malt milkshake. Lastly, said sandwiches cost from a piffling £8 with fries. Bish, bash, bosh.

7 St Stephen Street
(+44 131 225 8116,
www.bellsdineredinburgh.co.uk)

🔔 SPLASH OUT

Leith's the port district north of Edinburgh, and really a separate place in its own right. The converted warehouses are buzzing these days, revived by an art scene and hip restaurants. Foremost is **The Kitchin**, whose eponymous head chef Tom Kitchin took just six months to snaffle a Michelin star. This is Scottish "nature to plate" with a French twist: homemade Borders sausages with endive tatin and bramble sauce or roasted partridge in raisin-and-grape sauce. The six-course Chef's Surprise Tasting Menu offers a rangy showcase, while three-course set lunch deals offer very good value. Book well ahead.

78 Commercial Street, Leith
(+44 131 555 1755,
www.thekitchin.com)

▼ The Kitchin is located by the redeveloped waterfront in Leith, and is set in a former whisky storage building.

HIP & HAPPENING

Descend west from Arthur's Seat and you'll happen upon Edinburgh's most enterprising restaurant. *Aizle*—it rhymes with hazel—adheres to the bistronomie concept: formal food at informal prices in informal surroundings, with a short and seasonal carte. But even that doesn't fully explain the wonders that Stuart Ralston conjures up in this sallow, spartan space. Previous concoctions include skate with roasted parsley roots, miso-cured hake, pork terrine, sous-vide radishes and burnt apple sauce, and stout bread

▲ Aizle prides itself on making everything it serves in-house, using the best of local produce.

under whipped duck butter. Whatever the day's five courses, you'll find them handwritten on a wall board.

107–109 St Leonard's Street (+44 131 662 9349, www.aizle.co.uk)

LOCAL SECRET

Auld Reekie's not an obvious place to find high-quality Mexican food—although, actually, it has plenty—so most tourists tend to miss *El Cartel*, a sister establishment to neighboring bar-restaurant *Bon Vivant* in the elegant New Town. Switched-on locals sure don't, however, which explains the pocket-sized cantina's regular queues. Further explanation is provided by a no-reservations policy, and still more upon a first chowing of the homemade duck carnitas or fish tacos. And those lines seem positively practical by the time you scarf the mezcal cocktails and chocolate-and-chili ice cream.

64 Thistle Street
(+44 131 226 7171,
www.elcartelmexicana.co.uk)

¶◎¶ BRILL FOR BRUNCH

Available daily until 5pm—for which, bravo—brunch menus at **Urban Angel** have two sections: healthy breakfasts and (not always so healthy) cooked breakfasts. The former stars homemade granola, guilty pear, oatmeal porridge, and avocado with salsa, while its counterpart includes Arbroath smokies (i.e. smoked haddock) and lots of egg dishes. Combos are, of course, possible. Everything is sustainably sourced where possible, with lots of gluten- and dairy-free options. Come early on weekends, cozying up amid the wood fittings, and begin with a Breakfast Smoothie: almond milk, blueberries, banana, and honey.

121 Hanover Street
(+44 131 225 6215,
www.urban-angel.co.uk)

PUB GRUB
Edinburgh hardly lacks for pubs (and the quality is generally high), but few serve such legendary food as **The Cumberland Bar.** Most renowned are its Sunday roasts, and particularly the beef. You're advised to reserve at the best of times, and definitely when warm weather brings the lovely beer garden into play. Inside there's Black Isle Blonde on tap and bent-cane chairs described in Alexander McCall Smith's *44 Scotland Street.* Another New Town boozer with literary links is the simple, white-fronted **Oxford Bar**: it's to this

▲ The Oxford Bar, located in the New Town area of the city, has attracted local writers through its doors since the 19th century.

dark-wood den that Ian Rankin's DI Rebus regularly repairs, occupying church-pew stalls and drinking a pint of Deuchars IPA, brewed locally in the city.

1–3 Cumberland Street
(+44 131 558 3134, www.
cumberlandbar.co.uk); 8 Young Street (+44 131 539 7119, www.oxfordbar.co.uk)

 **REGIONAL COOKING**

As if the *Arcade Bar, Haggis & Whisky House*'s name, isn't evidence enough of its patriotic offerings, this Old Town pub near Waverley station has used Scotland's most famous poet to christen one plate. Order "Robert Burns' Famous Haggis" and you'll enjoy a tower of the spicy, minced sheep's pluck in, yes, whisky sauce. The chosen blend being just one of 100 varieties of Scotch on offer. Further up the Royal Mile, *The Royal McGregor* is equally flag-waving: its thistle wallpaper comes supported by various haggis contrivances, including the obligatory neeps and tatties (swede and potato).

▲ If you've never tried haggis, you must. Don't be put off by preconceptions—you won't be disappointed. Try it at the Arcade Bar.

48 Cockburn Street (+44 131 220 1297, www.arcadepub.co.uk); 154 High Street (+44 131 225 7064, www.royalmcgregor.co.uk)

KILLER COCKTAILS

In a medieval lane near St Giles' Cathedral, *The Devil's Advocate* complements its fancy food with clever cocktails—try a Build-Up, blending tequila, orange-and-ginger cordial, vermouth, and rhubarb bitters. A veteran of Edinburgh's mixology scene is *Bramble*, its cellar huddling beneath the New Town. There's much to like there: snug crannies,

▲ A haggis served with the traditional mashed neeps and tatties. Parsley garnish is a questionable extra.

tealights, weekend DJs, and out-there cocktails. The jingoistic Campbeltown (single-malt whisky, cherry liqueur, and chartreuse) is good, but nothing trumps a Rheingold Express (Bols Genever, cardamom-and-vanilla syrup, lemon juice, and crème de cacao).

9 Advocate's Close (+44 131 225 4465, www. devilsadvocateedinburgh. co.uk); 16A Queen Street (+44 131 226 6343, www.bramblebar.co.uk)

☕ CAFFEINE KICKS

Periodic-table tiles emphasize the geeky attention to detail on show at **Brew Lab Coffee**, found in Edinburgh's university heartlands near the National Museum. Specialty-grade beans rotate weekly, but all are high-quality and transformed into pour-over filters by equally impressive baristas. There's also a progression from the usual stripped-back, white-wall vibe found in modern coffee shops; distressed bricks and exposed floorboards still feature, but so do light bulbs hanging off red rope and slumpy crimson leather armchairs. Free wi-fi also encourages a longer stay, as do cakes and soups from local producers.

6–8 South College Street (+44 131 662 8963, www.brewlabcoffee.co.uk)

MARKET RESEARCH

You can find foodie bazaars for fun in Edinburgh now, but that wasn't true back in 2000. The launch of the matter-of-factly-named **Edinburgh Farmer's Market** (Sat, 9am–2pm) changed everything, and this pioneer is still going strong today. Spectacularly held under the castle, it's best known for lunchtime buffalo burgers and hog-roast rolls. Slightly smaller and less central is **Stockbridge Market**

(Sun, 10am–5pm). If you don't go there for the Steampunk Coffee, artisan patisseries, and culinary crafts, go for the location: in the lovely Jubilee Gardens and beside the Water of Leith, which you can stroll along afterward.

Castle Terrace (+44 131 220 8580, www.edinburghfarmersmarket. co.uk); 1 Saunders Street (+44 131 261 6181, www.stockbridgemarket.com)

▼ Scottish produce is famed for its impressive quality, much of which is to be found at Edinburgh Farmer's Market.

London

ENGLAND

> London's an epicurean beast. It has hotspot culinary neighborhoods—Marylebone, King's Cross, Victoria—which each merit an individual guide. It has as many Michelin stars, street-food fairs, gastropubs, and pop-ups as red buses, plus perhaps the world's best coffee. It even has Gordon "f*cking" Ramsay. But where to start?

HIP & HAPPENING Brixton makes a worthy rival, but Peckham is really South London's coolest area: its trendies and train-arch bars are accompanied by an authentic, multi-ethnic vibe. *Voodoo Ray*'s pizza is mighty popular, and *Artusi* serves delicious Mediterranean dishes, but the hottest foodie ticket is *Taco Queen*. Now installed in a tiny, permanent space on main drag Rye Lane, it accompanies delicious Baja fish tacos and jackfruit arepas

with craft beer and the friendliest service imaginable. Have a cocktail beforehand at *Frank's Café* (summer-only), atop a former car park, and finish the night dancing to soul jams in the *CLF Art Café*.

191 Rye Lane (no phone, www.twitter.com/tacoqueenldn)

💲 ON A BUDGET?

A common abbreviation in London, BYO stands for "bring your own" and most often refers to licensed Asian restaurants that allow patrons to arrive with alcohol, keeping down already-low prices. Most famous in this bracket are the *Mien Tay* restaurants, a Vietnamese chain found across London. The pair just north of Shoreditch are perfect for East London explorers, and usually have free tables. Try chargrilled quail with honey and garlic, papaya salad, or the very dependable sliced-beef pho. BYO is available Sunday–Thursday, with a £2.50 corkage charge.

106–108 or 122 Kingsland Road (+44 20 7729 3074 or +44 20 7739 3841, www.mientay.co.uk)

🔔 SPLASH OUT

You'll spend over £100 for three courses at *Restaurant Gordon Ramsay*, but it's well worthwhile for the showcase of French cuisine using the best English ingredients: Cornish brown crab with lovage and lemon thyme, say, or pigeon beside beetroot and quince. You'll soon understand why the place boasts three Michelin stars. A bit more relaxed is *Ceviche* in Soho. Martin Morales's don ceviche is the premium order there, thanks to its marvelous sea-bass slices marinated in thrilling leche de tigre sauce, but virtually every dish boasts tangy thrills. The pisco sours are worth a visit alone.

68 Royal Hospital Road (+44 20 7352 4441, www.gordonramsayrestaurants.com); 17 Frith Street (+44 20 7292 2040, www.cevicheuk.com/soho)

▼ Restaurant Gordon Ramsay is the foundation upon which all the chef's other successes are built.

LOCAL SECRET

Unless attending the nearby tennis tournament, you're unlikely to end up in South Wimbledon. Few people ever end up in South Wimbledon. Do so, though and you can frequent one of London's greatest Japanese restaurants. At minimal *Takahashi* (Wed–Sun only), an ex-Nobu alumni cooks wonderful wasabi-infiltrated ceviche, scallops beside palate-thrilling citrus mayo, the supplest snow-crab nigiri imaginable, and so on. His wagyu beef is so sensational that you'll be tempted to book a trip to Japan on the spot. There are good cocktails, too, and beautiful glazed plates, and fabulous service. It's pretty good, this South Wimbledon.

228 Merton Road (+44 20 8540 3041, www.takahashi-restaurant.co.uk)

BRILL FOR BRUNCH

London's most popular brunches are at *The Breakfast Club*'s various branches, outside of which lines invariably snake. Are they worth the effort? It's debatable. Better, then, to head to *Caravan*'s outpost in King's Cross station's rehabilitated goods yard, where university students and foodies have replaced the bags of wheat and faulty freight trains. Beneath big pipes in a spacious, exposed-brick room, Caravan serves specialty coffee and a huge brunch menu. The options extend from various granolas and coconut bread with lemon-curd cream cheese to brownies or blood-sausage hash and poached eggs.

1 Granary Square (+44 20 7101 7661, www.caravanrestaurants.co.uk/kings-cross.html)

QUINTESSENTIALLY LONDON

It doesn't get more London than an indulgent afternoon tea. Despite numerous imitators, Mayfair's art-deco *Claridge's* hotel still serves the ultimate spread. Usually accompanied by a harpist, this constitutes tiered silver trays of finger sandwiches, warm scones with clotted cream, and more cakes than your dentist ever need know about, plus posh teas. Another English classic is Sunday roasts, where slices of meat come with "all the trimmings": roast potatoes, various veg, and, if it's beef, a Yorkshire pudding. Try *The Harwood Arms*, a cozy, Michelin-starred "posh pub" on the Chelsea/Fulham border.

41–43 Brook Street (+44 20 7629 8860, www.claridges.co.uk); 27 Walham Grove (+44 20 7386 1847, www.harwoodarms.com)

▼ Situated on a square that's perfect for people-watching, Caravan is highly recommended for al fresco dining.

CARAVAN
KING'S CROSS

🫕 REGIONAL COOKING

If it's straight-up British dishes you want, aim for semi-formal *St. John* in a former Smithfield Market smokehouse. Chef Fergus Henderson is renowned for his pigeon pies and wigeon legs; for his smoked eels and Welsh rarebits. The dessert menu is just as jingoistic, headlined by apple crumbles and bread pudding—served with hot butterscotch sauce. One thing you won't find is fish and chips. For that, go west to Marylebone and the *Golden Hind*, in among the trending neo-bistros. Battered cods have been sold there for over 100 years.

26 St. John Street
(+44 20 7251 0848,
www.stjohngroup.uk.com/
smithfield);
71a–73 Marylebone Lane
(+44 20 7486 3644, www.
goldenhindrestaurant.com)

▲ The Next Big Thing comes and goes, but St. John continues to wow foodies with dishes inspired by its nose-to-tail ethos.

🍸 KILLER COCKTAILS

Where to start? From Chinatown's chilled *Experimental Cocktail Club* to Shoreditch's speakeasy-style *Happiness Forget* to plush, award-winning *Dandelyan* on the South Bank, London abounds with legendary mixology haunts. One of the most under-rated, however, is *The Four Sisters Bar* in well-to-do Islington, just off Upper Street. You won't see show-off juggling or paper umbrellas there; instead there'll be homemade syrups lined up like an apothecary's, barrel-aging negronis, and lots of romantic candles. The affable bartenders can make perfectly any drink you can name, and many that you can't. Visit on school nights to ensure a seat.

---

25 Canonbury Lane
(+44 020 7226 0955,
www.thefoursistersbar.co.uk)

☕ CAFFEINE KICKS

Getting a good cup of coffee is easy in London. If it's coziness you're after, try *Esters* on the backstreets of chichi Stoke Newington. Otherwise, make for Fitzrovia, the city's prime java hub. There's *Attendant*, in an old toilet, a branch of *Workshop*, another of *TAP*, and, best of all, the two locations of *Kaffeine*. Their original place has revived its guest espresso

▼ Coffee served from a Victorian toilet might not sound appealing, but a flat white from Attendant will prove you wrong.

program, introducing a glittering rota of global roasters, and offers high-tech gadgetry and chicly minimal but cozy decor. If tea's your poison, meanwhile, head a few blocks east and visit specialist café *Yumchaa*.

66 Great Titchfield Street (+44 20 7580 6755, www.kaffeine.co.uk); 9–11 Tottenham Street (+44 20 7209 9641, www.yumchaa.com)

 **MARKET RESEARCH**

London's oldest market—at eight centuries and counting— is also its best for food. Held on a sprawling site below viaducts

▲ It might now be overrun with tourists, but Borough is still far and away London's best food market.

near the southern end of London Bridge, *Borough Market* (Mon–Sat, 10am–5pm; limited stalls, Mon and Tues) is distinctly gourmet. You can come for souvenirs such as blackberry jams or extra-virgin olive oil. Or you can come to eat, with venison burgers, chickpea dosas, and wild-boar tortellini among the available fodder. Eastward along Tooley Street is *Maltby Street Market* (Sat, 9am–4pm; Sun, 11am–4pm), a smaller, much narrower affair. Look out for artisan brownies, gin, raclette, donuts, and an offshoot of St. John.

8 Southwark Street (+44 20 7407 1002, www.boroughmarket.org.uk); 41 Maltby Street (no phone, www.maltby.st)

Ghent

BELGIUM

> Ghent's canals, cathedrals, and gabled buildings—all beautifully illuminated at night—provide the backdrop for some seriously good, seriously under-rated eating. This is a hub for the Flemish Foodies, three young chefs ignoring the rule book, and Europe's best city for vegetarians. Also present is some typically excellent Belgian beer.

SPLASH OUT

The most famous Foodie is Kobe Desramaults, who has recently closed his much-loved De Vitrine restaurant and replaced it with *Chambre Séparée*, using the disused Belgacomtoren skyscraper until 2020. There's no menu: you simply enjoy the 20-course tasting menu on which Desramaults has settled that day. Only 16 spaces are available, and everyone sits facing an open kitchen. Reservations must be made months ahead—groups of two maximum—and it'll cost you €205, plus alcohol. Alternatively, try *Cochon de Luxe*, where tiny plates marry gastronomy with art, humor, and, counter-intuitively, a seedy Red Light District location.

Keizer Karelstraat 1 (no phone, www.chambreseparee.be); Brabantdam 113 (+32 9 336 1672, www.cochondeluxe.be)

HIP & HAPPENING

Alain Coumont founded Le Pain Quotidien, so he clearly knows his stuff. *Le Botaniste* is Coumont's centrally located contribution to Ghent's claim to being the "Vegetarian Capital of Europe." Patrons order, pay, wait, and collect at the café, then take their tray to a table stool, the tiny terrace, or a back conservatory, admiring nice checkerboard flooring en route. The food's all organic and nourishing fare, such as coconut and peanut-butter curry with black rice and steamed spinach. Coumont is also a vigneron; hence the array of wines on shelves. If possible, visit on a Thursday—the city's appointed "Veggie Day."

Hoornstraat 13 (+32 9 233 4535, www.lebotaniste.be)

▲ Thanks to the efforts of local chefs, Ghent has forged itself a reputation as a destination for some exceptional, forward-thinking vegetarian cooking.

💲 ON A BUDGET?

Linked to all the vegetarianism, soup kitchens are also a common sight across Ghent, and ideal for anyone watching their Euros. One of the best is located between the River Leie and Confiserie Temmerman candy store. Tomato, pumpkin, and broccoli soups are available daily in *Souplounge* and cost €5 for large portions served with two bread rolls, butter, and some fruit. Toppings—cheese, croutons, meatballs—are only a little extra. The pale-wood *SOUP'R*, found just off the central Korenmarkt, is another cooler, cut-price option.

Zuivelbrugstraat 4 (+32 9 223 6203, www.souplounge.be)

💬 LOCAL SECRET

To find *La Malcontenta*, you must leave riverside Kraanlei and dive down a residential, lantern-lit lane: it looks lovely, but decidedly unpromising. After a hundred yards, you turn onto another identikit street, and suddenly there it is: a beckoning neighborhood tapas bar emitting yellowy light and a hum of happy chatter. In other words, this isn't somewhere you'll happen on by accident. Those in the know enjoy unusual homemade tapas such as saffron rice with cranberries, Canary Island wines, draught beers from a small bar, and cumin-spiked bread. You can sit inside or on a flower-filled back terrace.

Haringsteeg 7 (+32 9 224 1801, www.facebook.com/tapasbarlamalcontenta)

🍽️ BRILL FOR BRUNCH

Narrow streets and higgledy houses make Patershol the city's most atmospheric quarter. It's home to Ghent's best breakfast option, too, in British-run B&B/café *Simon Says*. Housed in a dazzling Art Nouveau building, the two rooms contain a mural by avant-garde artist Panamarenko and—of far more importance to foodies—a Faema E61 coffee machine. Food-wise, you've got toasts, eggs, and croques to choose from. Another option is *De Superette* (Thurs–Sun only), overseen by Kobe Desramaults and housing thrift furniture inside a former grocery store. But while its coffee, freshly baked breads, and atmosphere impress, the food's a bit hit-and-miss.

Sluizeken 8 (+32 9 233 0343, www.simon-says.be)

REGIONAL COOKING

Mémé Gusta has television origins: founders Nele Smet and Jan Hendrickx created it for the *My Pop-Up Restaurant* show, and then went on to open this cozy haunt with the same concept—namely, pure and nostalgic Flemish cuisine, delivered in a no-holds-barred fashion. Meme translates as "Grandma" for this very reason. You can order beery stews, boiled sausage on a heap of turnip mash, or a vol-au-vent of creamy chicken and mushroom. The mish-mash decor—wooden tennis rackets, a chandelier, vintage cans—makes good use of Nele's nearby used furniture and accessories business.

Burgstraat 19 (+32 9 398 2393, www.meme-gusta.be)

 ## QUINTESSENTIALLY GHENT

Another Foodie is Jason Blanckaert, who gave up a Michelin star at his chic C-Jean to open a more casual bistro in Patershol—a move synonymous with Ghent's history of rebellion and free-spiritedness. *J.E.F* is almost pointedly plain: a small square room beside a lovely old tree. Blanckaert describes his newest adventure as "real food," and you can easily see why: slow-cooked veals, oven-braised cods, and lusty pork bellies are the wholesome order of the day, ably supported by 16 quality bottled beers. Three lunch courses cost a paltry €28.

Lange Steenstraat 10 (+32 9 336 8058, www.jefgent.be)

▽ KILLER COCKTAILS

Ghent's former postal sorting office was built with 1913's World Fair in mind; hence the Gothic spires and general magnificence. Its top two floors have recently reopened as the *1898 The Post* hotel, a luxurious dream of demure interiors and high ceilings. There too is *The Cobbler*, a cocktail bar open to non-guests. First, enjoy the homely layout: wooden floors, textured wallpaper, and lots of carefully sourced, fin-de-siècle clutter. Then focus on classic and new-invention potions like The Cobbler Negroni, delivered in a chrome goblet beneath fresh berries. Plates of the best local cheese and charcuterie can be enjoyed alongside drinks.

Graslei 16 (+32 9 273 9020, www.zannierhotels.com/1898thepost/experience/the-cobbler)

▼ A large student population and a reputation as a cultural hub has ensured Ghent maintains a healthy bar scene.

CAFFEINE KICKS

South of the Ketelvest canal begins Ghent's university area, although lively Walpoortstraat also possesses some chichi shops—including Yuzu, one of the world's great chocolatiers—and the counter-cultural arts venue Vooruit. Watch all the action and eclectic footfall from *OR Coffee Shop*, a sprawling and welcoming café whose seating options span cutesy window seats and many a sofa. Headphoned students clack away on MacBooks, as friends catch up over a premium espresso or slice of strawberry tart. There's another branch in northern Ghent, and two in Brussels, but none rival this haunt's relaxing vibe.

Walpoortstraat 26 (+32 9 223 6500, www.orcoffee.be)

▲ The Waterhuis aan de Bierkant stocks a vast selection of beers—and matching glassware—that's second to none.

BEER OH BEER

At night, lights dance on the River Leie's surface, as canoeists glide toward St Michael's Church with candles in their bows, and the wonky, pretty guildhouses glow a honey color. It's a lovely scene, and best appreciated from outside the *Waterhuis aan de Bierkant*. You can order numerous beers there, from Trappist blondes to sour red Rodenbach. The waterside seats are shared with Flemish restaurant *Chez Leontine* and the dingy *Dreupelkot* "brown bar," whose famously grumpy owner, Pol, stocks 150 brands of jenever, or gin, headed by a homemade vanilla variety. So, basically, you can quite contentedly sit there for hours.

Groentenmarkt 9 (+32 9 225 0680, www. waterhuisaandebierkant.be/97)

Amsterdam

THE NETHERLANDS

As little as a decade ago, Amsterdam's chances of making this book would have been slim. But an intense boom of neo-bistros, street food, locavorism, and of, well, everything has transformed things. Once a place where you tolerated the food, the Dutch capital now has restaurants to relish.

REGIONAL COOKING

While there's no singular Amsterdammish cuisine, Dutch produce is being increasingly championed. Take *Rijks*, the restaurant of the neighboring art- and history-focused Rijksmuseum, where 30-something Joris Bijdendijk cooks wholesome, unshowy small plates of domestic inspiration. Think beef cheek and apple foam; duck with sweet-and-sour fennel; spit-roasted celeriac beside messenklever cheese; halibut and chorizo; herrings under lemon-and-champagne chutney. Dutch spirits— such as local favorite jenever, a precursor to gin—and wines are also incorporated where possible. The space is aptly chamber-like: high ceilings and crisp acoustics.

Museumstraat 2 (+31 20 674 7555, www.rijksrestaurant.nl)

QUINTESSENTIALLY AMSTERDAM

So many Amsterdam establishments are enterprising, be it by repurposing unlikely venues or serving up eclectic cuisine. Typical of all that is *Instock*, which became the country's debut "food-waste" restaurant a few years back. The venture was a success, and two further branches in Utrecht and The Hague soon followed. The original, now located in the Eastern Docklands, continues to make innovative use of whatever store leftovers it "rescues" that morning, so making a point about society's profligacy. A very delicious point, especially where the pork and carrot dumplings are concerned.

Czaar Peterstraat 21 (+31 20 363 5765, www.instock.nl)

ON A BUDGET?

Amsterdam ain't cheap, with quality feeds for less than €20 being hard to find. One place offering such a rarity is *Moksi* in the hipster village of De Pijp. Its basic, chaotic interior belies superior Surinamese-inspired cooking (Surinam was a former Dutch colony): lamb masala curries, wild duck roti, shrimps steeped in tomato sauce. The opening hours aren't always upheld, but the risk's easily worthwhile. Alternatives include most joints in Chinatown, or *La Perla* in canal-lined Jordaan, whose margherita or porchetta wood-fired pizzas start at €9.50. There's homemade ravioli for lunch, too.

Ferdinand Bolstraat 21 (+31 20 676 0264, www.moksi.nl); Tweede Tuindwarsstraat 53 (+31 20 624 8828, www.pizzaperla.nl)

▼ Instock represents a socially conscious approach to food that never has to sacrifice on flavor.

HIP & HAPPENING

While inevitably gentrifying, Amsterdam-Noord—which is reached via free ferry from Centraal Station across the River IJ—remains a post-industrial playground for street artists, skateboarders, and beatniks. In one ex-shipyard is *Pllek*, whose reclaimed petrol containers and ivy-snagged scaffold pipes give onto a sofa-speckled, deckchair-dotted veranda and sandy beach for urban swimmers. It's a sprawling and deeply relaxing spot, to which you can go for coffee, yoga, live music, exhibitions, or organic nosh such as steamed mackerel and horseradish crème fraîche. A little to the east is *Café De Ceuvel*, where upcycled houseboat units have reclaimed previously polluted grounds.

TT Neveritaweg 59 (+31 20 290 0020, www.pllek.nl)

SPLASH OUT

It's far easier now to spend over €50, as luxury hotels mushroom and Michelin stars flutter about. *Vermeer* boasts only one, but that's on the judges' consciences: British chef Chris Naylor's piquant, greens-focused cuisine—many of the ingredients hail from a rooftop garden—is outstanding. Menus alter with the weather, but might include cider-marinated mackerel ceviche and cucumber ice cream or duck breast plus beetroot risotto. Like Naylor's approach, *Vermeer* has enjoyed a recent makeover, swapping white tablecloths and chandeliers for exposed, bruised walls and leaf-swamped black shelving.

▼ You can turn up at Pllek for breakfast and find plenty going on to keep you busy until the early hours of the following day.

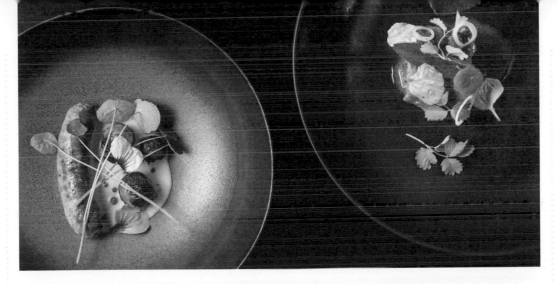

▲ Restaurants like Rijks (see page 36) have helped Amsterdam to develop food that's contemporary but rooted in Dutch tradition.

Prins Hendrikkade 59–72
(+31 20 556 4885,
www.restaurantvermeer.nl)

💬 LOCAL SECRET
Well-to-do, partly residential, and business-focused, South Amsterdam is an unlikely part of town to find one of the city's best dining spots; ditto the Brutalist ex-chapel occupied by *Restaurant As* with its surrounding terrace and cherry trees. Override those instincts, however, as the food served there is magnificent: every last locally sourced carrot or cob bursts with organic flavor. Supported by independently produced wines, the short, bistro-style menu alters daily, and could just as well deliver orecchiette under tomato and anchovy as porchetta with pumpkin, chanterelle mushrooms, and celery sauce. It matters not: everything is outstanding.

Prinses Irenestraat 19
(+31 20 644 0100,
www.restaurantas.nl)

🍴🍽️ BRILL FOR BRUNCH
As well as Moksi, De Pijp's many establishments extend to *Little Collins.* In tribute to the Aussie owners' Melbourne home, this cozy rectangle has turned brunch—available daily except, bizarrely, on Tuesdays—into a science. On the international menu? Corn fritters with guacamole. Bangers and mash. Coconut-coated French toast with l emon curd. Sweet-and-sour pork belly with Asian 'slaw. Accompanying mocktails and cocktails, including various spicy Bloody Marys, help wash it all down. Dinners are also offered from Wednesday through to Sunday, but late morning is the standout time to indulge.

Eerste Sweelinckstraat 19F
(+31 20 753 9636,
www.littlecollins.nl)

KILLER COCKTAILS

Love a good view? Then ride up the city's iconic A'DAM Tower to *Madam*, its 20th-floor bar. The glorious vistas are complemented by live DJs, good food, and a kooky cocktail list. Need freshening up? Try a Dutch Spring—jenever, honey, and lime. Up for partying? It'll have to be a Hypebeast (Bacardi, maple syrup, and pecans). The classics are also available, along with local lager on draught. Across in Chinatown, *Dum Dum Palace* doesn't restrict itself to pan-Asian food;

▲ Sleekly contemporary, Madam offers cutting-edge mixology alongside a panoramic view of the city.

you can also order oriental potions like the Matcha Mule.

Overhoeksplein 3 (+31 20 237 6310, www.madamamsterdam.nl); Zeedijk 37 (+31 20 304 4966, www.dumdum.nl)

CAFFEINE KICKS

Sick of starkly white-walled coffee shops, with stencil-lettered menu boards and tiny stools? There are plenty of those in Amsterdam, but *Sweet Cup* is refreshingly different. Despite the alabaster exterior, all's lusher and cozier inside: houseplants splay out of chunky pots and walls see planks of different muted colors—pistachio, cocoa powder, and honeycomb—

patchworked around an L-shaped bar and, hooray, proper seats. The company's own-roasted beans are used, translated into delectable liquid courtesy of a Synesso machine. Usually overseeing it all is pet basset hound Sjefke.

Lange Leidsedwarsstraat 93HS (+31 20 370 3783, www.sweetcupcafe.com)

MARKET RESEARCH

Stroopwafels are cookie-like sandwiches with a caramel filling. You will find them throughout the Netherlands, but they are best bought in Amsterdam at the *Albert Cuyp Market*, or more simply, ***The Cuyp*** (Mon–Sat, from 9:30am). This century-old street affair, in De Pijp, also stocks fruit, veg, meat, and other inedible souvenirs such as jewelry. Over in trendy Oud-West, the ***Foodhallen*** (daily, 11am–11:30pm) is a grand space in which one can eat Vietnamese spring rolls, grilled cheese sandwiches, bitterballen (beef and veal balls), white-bean falafel wraps, and lots more inside a former tram depot.

Albert Cuypstraat (no phone, www.albertcuyp-markt. amsterdam); Bellamyplein 51 (no phone, www.foodhallen.nl)

▼ With street food, craft beer, cocktails, and DJs, Foodhallen is a popular destination for Amsterdam's cool crowd.

Berlin

GERMANY

> The food-and-drink scene in Germany's capital has improved so markedly that it now rivals clubbing as a reason to visit. Like the nightlife, victuals there tend to be staunchly Bohemian in nature; there's a lot of meat, an overgrown Thai picnic, and even a bar where Goebbels once committed adultery. How very Berlin.

$ ON A BUDGET?
Friedrichshain's *Pizza Dach* has something of a bad reputation, for being a) cheap, b) open 24/7, and c) on a lot of nighthawks' daybreak route from nearby club Berghain. Yet, despite the characterization as a drunken eat, it ably than stands up to sober daytime inspection. Margaritas cost €3.50, and toppings—salami, olives, corn—are a few cents extra. For a more historic form of low-cost Berliner culture, visit an imbiss (snack bar) for some currywurst.
Konnopke's Imbiß on Schönhauser Allee

▲ Konnopke's Imbiß has been satisfying local wurst cravings for over 80 years.

▼ The Deutsches Currywurst Museum (yes, that is a real place) estimate that Berliners eat 70 million currywurst each year.

is perhaps Berlin's most famous, but the best one is *Krasselt's*, a fair schlep to the south-west in Steglitz.

Wühlischstraße 32 (+49 30 2904 4117, no website); Steglitzer Damm 22 (+49 30 796 9147, www.krasselts-berlin.de)

SPLASH OUT
It might (bafflingly, ludicrously) lack a Michelin star, but *Glass* is still Berlin's best dining experience. Inside a Brutalist West Berlin apartment block, 30 something Israeli chef Gal Ben-Moshe conjures up a winning blend of playfulness, Heston-style molecular experimentation, tastebud-challenging sweet/savory combos, and modern Arabic cooking. In his relaxed, black-walled den you might be served duck with coffee and chocolate or carrot soup and baklava. The Candybox dessert, its "Snickers snow" and passionfruit gummy bears recalling childhood candy, is legendary. There are five-, seven-, or nine-course tasting menus to choose from.

Uhlandstraße 195 (+49 30 5471 0861, www.glassberlin.de)

BEER OH BEER

Open-air biergärten (beer gardens) are a Berlin staple, and often the first stop on summer nights out. For beauty and class, head to Tiergarten park, where *Café am Neuen See* serves seasonal food, plus snacky pretzels all year beside a boating lake. Come the warmer months, however, the real hotspot is leafy *Schleusenkrug*, a self-service beer garden just a short 10-minute walk to the west of the park. Over in north-east Berlin's Prenzlauer Berg, *Pratergarten* is the city's oldest beer garden. Various

▲ Schleusenkrug really comes into its own in the hot summer months, providing refreshing cold lagers to combat the heat.

local-brewed pils are served under its lovely chestnut trees between May and September, along with some food.

Müller-Breslau-Straße 1 (+49 30 313 9909, www.schleusenkrug.de); Kastanienallee 7–9 (+49 30 448 5688, www.pratergarten.de)

LOCAL SECRET

While cult *Mustafa's* tends to attract Berlin's kebab headlines, there's plenty more shish in the sea in this most carnivorous of destinations. Much more under the radar—and, therefore, infinitely less busy, even allowing for 10-minute waits at peak time—is *Rüya Gemüse Kebab* in the southern district of Schöneberg. Seriously delicious doners there come in gently crisped flatbread, attended by sprigs of fresh herbs and just-grilled vegetables. You can even chuck in some crumbled feta. Yes, it's out of the way, but you won't regret the trip.

Hauptstraße 133 (+49 30 6956 4413, www.facebook.com/ruyagemusekebab)

🍴 BRILL FOR BRUNCH

Found down a backstreet in uber-cool Neukölln, *Roamers* is filled with pot plants and knick-knacks—a nice, homely kind of clutter. It's also filled with customers, by virtue of only seating around 20 people, so try to arrive early (you can't reserve). Giant plates of poached eggs and spinach are a common order, along with French toast plus crème fraîche

▼ A healthy way to start the day at Roamers.

▲ Vibrantly purple in color, borsch is a beetroot soup that has fueled the Prussian population for centuries.

or Greek yogurt, grilled pears, and berries. If you're based nearer Friedrichshain, try *Silo Coffee* for its baked eggs in feta-sprinkled tomato sauce. Do you want that with chorizo? Of course you do!

Pannierstraße 64 (no phone, www.roamers.cc); Gabriel-Max Straße 4 (no phone, www.facebook.com/silocoffee)

 REGIONAL COOKING

If it's the old Germany you want, make for *Marjellchen*. Not only are there vintage watercolors on the walls and old folk music crackling from long-players, but Prussian classics dominate the time-warp menu, too. You can order tempestuous borsch or Königsberger Klopse—meatballs

in a white-wine sauce of capers and anchovies—in generous portions, plus the obligatory lager. Over in the district of Mitte, bright white *Lokal* is firmly routed in the present day, meanwhile, what with its embracing of nose-to-tail and locavore philosophies. Pork belly with fresh radishes and turnips is a typical dish.

Mommsenstraße 9 (+49 30 883 2676, www.marjellchen-berlin.de); Linienstraße 160 (+49 30 2844 9500, www.lokal-berlin.blogspot.co.uk)

QUINTESSENTIALLY BERLIN

A common feature of Berlin is its hidden courtyards, many containing semi-hidden bars or restaurants. Gently Bohemian in feel is *Kanaan*, a converted scrapyard hut turned Israeli eatery in Prenzlauer Berg, its balcony and urban garden facing railway sidings. The hamshuka (hummus and shakshuka combined) is renowned, as are the malawach—filled Yemenite pancakes with mango curry sauce. Enjoy *Kanaan* and you'll probably like *Klunkerkranich*, too. This time the venue is a shopping-mall rooftop in deepest Neukölln, where party animals can eat street food, booze, and party all night.

▼ People relax during the day in Klunkerkranich's rooftop garden, with views of the city center to the north.

Kopenhagenerstraße 17 (+49 176 2258 6673, www.kanaan-berlin.de); Karl-Marx-Straße 66 (no phone, www.klunkerkranich.de)

🍸 KILLER COCKTAILS

None of Berlin's fabled cocktail bars are quite as surreal as the dim and diminutive *Rum Trader* in unassuming Wilmersdorf. Forget menus; there, you just tell the bartender what you'd like, then watch them whizz up something mind-bogglingly unusual and delectable. No biggie. Locals chain-smoke, jazz plays softly, and phones seem a distant memory. Ditto in *Bar Lebensstern*, a grand, 19th-century Schöneberg villa where Joseph Goebbels once installed his mistress. Today the various red rooms host barflies, many of whom sip ranglums (lots of dark rum, less white rum, falernum, and lime).

Fasanenstraße 40 (+49 30 881 1428, no website); Kurfürstenstraße 58 (+49 30 2639 1918, www.cafeeinstein. com/lebensstern-cocktailbar)

☕ CAFFEINE KICKS

What's your preference: extremely high-quality, fastidiously made, third-wave coffee or simply pretty good stuff in a lazy, relaxing space? If it's the former, then Mitte's *The Barn Café* is the place to go (with polite reference to *Bonanza* and its two high-design stores). There you'll find four different brew methods, from Aeropress to Syphon, reverse-osmosis filtered water, and a regularly changing menu of direct-sourced roasters. Just don't mention soy milk or open a laptop. Both of which practices are entirely acceptable in *Leuchtstoff*. This laid-back Neukölln retreat is full of beckoning crannies, couches, and, yes, plug sockets.

Auguststraße 58 (+49 151 2410 5136, www.thebarn.de); Siegfriedstraße 19 (+49 177 196 1512, www.facebook.com/ leuchtstoffK)

**MARKET
RESEARCH**

There are street-food fairs and
then there's Street Food
Thursday inside Kreuzberg's
lovely *Markthalle Neun*.
Some 5,000 diners turn up to
this truly global bash from 5pm
on the appointed weekday to
try Taiwanese bao burgers,
New Zealand-style pies, and
much more. The same space
also hosts an unusual Breakfast
Market on the third Sunday of
each month. Talking of unusual,
Wilmersdorf's Preußenpark

hosts the *Thaipark* on
summer weekends. Begun
two decades back as a few
Thai families gathered for a
grand picnic, this event now
sees its papaya salads and
shaved ice desserts lure large
local crowds.

Eisenbahnstraße 42/43
(ı 49 30 6107 3473,
www.markthalleneun.de/
street-food-thursday);
Fehrbelliner Platz (no phone
number, www.thaipark.de)

▲ At Thaipark you will find
some of the most authentic
Thai street food served in
the city.

Warsaw

POLAND

> Poland's capital can satisfy the most discerning of stomachs. From Communist-era milk bars gone trendy to cool coffee shops and repatriated market halls, the foodscape there is rich and rangy. As far as districts go, central Sródmiescie is the main, flower-happy hub, but slightly outlying areas such as Muranow and Mokotów also hoard gastro treasure…

💲 ON A BUDGET?

Poland is generally bargain-priced for most western travelers, so watching the pennies oughtn't be necessary. But one place to keep costs truly low is *Falafel Bejrut*, located near the eye-catching POLIN Museum of the History of Polish Jews, a little northwest of the center. Part of Warsaw's resurgent vegetarian scene, it offers hearty breads with combinations of turnips, tomatoes, peppers, mint, and more, plus ample hummus, some baklava, and vegan craft ice cream in a smart, simple interior. If you try really, really hard, you might just spend 50 zloty.

Nowolipki 15 (+48 50 151 2965, www.facebook.com/falafelbejrutwarsaw)

🛎 SPLASH OUT

Warsaw doesn't lack for fine dining, with plenty of Michelin stars on show. *Atelier Amaro* and *Senses* are two of the standouts, but neither rival *Nolita* for, well, sexiness. Compact, central, elegant, and decidedly modern—thanks to demure lighting and jet-black crockery—it's home to chef Jacek Grochowina, previously of London's Ritz, among other places. His beautifully presented à-la-carte offerings merge together Polish and international cuisine, and might include boiled octopus and lobster terrine, port-marinated foie gras, truffle mousse-stuffed quail, and deer tenderloin with pumpkin and smoked plums.

Wilcza 46 (+48 22 292 0424, www.nolita.pl)

LOCAL SECRET

How come *Flaming & Co*'s so popular? It could be the cheery white clapboard and charming, garden-style outdoor seating (the business also houses a florists). Perhaps it's all the fresh produce used in daily-changing menus, which draw from Italy, France, and the USA. Maybe it's the parkside location? How about the burgers? Or the bagels? The likeliest candidates, however—given the sheer number of people scoffing them—are some superb pizzas. Alongside classics like Bufala mozzarella and prosciutto await some stranger compositions, such as salmon sashimi or pumpkin and cranberry.

Chopina 5 (+48 22 628 8140, www.flaming-co.com)

 HIP & HAPPENING

Soho Factory is a 20-acre, post-industrial space given over to creative enterprises, one which aims to have the "spirit of Warhol" pervading. One of its former factories now houses *Warszawa Wschodnia*, a sort of informal fine-dining mash-up of French and Polish cuisines; examples include duck dumplings and stewed burbot in crab sauce. The red-brick interiors and large windows make this a particularly pleasant hangout, while the open kitchen grabs eyes and occasional piano recitals please ears. Go for lunch on a weekday and it's just 25 zloty for three courses, too.

▼ Warszawa Wschodnia has an open kitchen where you can watch the chefs at work. It's also open 24 hours a day, seven days a week.

Mińska 25 (+48 22 870 2918, www.mateuszgessler.com.pl/restaurants/warszawa-wschodnia)

 **BRILL FOR BREAKFAST**

Magda Gessler is Poland's culinary queen: a TV chef, foodie game-show judge, occasional film star, and restaurateur. One of her Warsaw outposts is the two-level *Slodki...Slony* ("Sweet...Salty"). Run the ground-floor gauntlet of some sumptuous-looking desserts—pink meringues, raspberry tarts, rose buns—and you'll find an upstairs café slash savory sanctuary where quality salads, soups, and omelets are served. Of course, there's nothing to stop you finishing with a little dessert… For a comfier brunch, head to bookstore-café *Kawiarnia Kafka* for pancakes in a student-popular space or deckchair-dotted garden.

Mokotowska 45 (+48 22 622 4934, www.slodkislony.pl); Obożna 3 (+48 22 826 0822, www.kawiarnia-kafka.pl)

🍲 REGIONAL COOKING

Want to try some traditional Polish dishes? There's no better place than **Stary Dom**, whose chefs adapt old-school meals to the 21st century. The options include fatty *golonka* (pork knuckles), *kotlet schabowy* (flattened veal schnitzels), platters of Polish charcuterie, and wild boar, duck, veal, or pork pâtés. If you've noticed the common theme there, well done; meat does indeed predominate. Other reasons to go include homeliness courtesy of high-beamed ceilings and low lights, the family-run vibe, and an adventure playground for juniors. Best of all, there's a distinct sense of authenticity in somewhere which could easily be a tourist trap.

Puławska 104–106
(+48 22 646 4208,
www.restauracjastarydom.pl)

📋 QUINTESSENTIALLY WARSAW

Poland's canteen-style milk bars are over a century old, but reached a peak of popularity during the 1950s. The main features were big portions of regional staples, low prices, plain-to-shabby decor, and a fearful, matron-like serving lady. Many have now moved onto the big restaurant in the

▲ Carnivores will feel right at home at Stary Dom, where meat dishes predominate on the traditional menu.

sky, but a few survivors are prospering in retro glory. Bar Bambino is Warsaw's most famous, but fame means crowds, so head instead to **Bar Ząbkowski**. Order carrot salad or cabbage and mushroom dumplings, and drink in the nostalgia of Communist-era dining for he disadvantaged.

Ząbkowska 2
(+48 22 619 1388,
www.barzabkowski.waw.pl)

🍺 BEER OH BEER

Another recurrent Warsaw feature is the city's grandma-style pub-bars: shabby, eclectic, and utterly informal dens where anything could—and does—happen. Translating, promisingly, as "In the clouds of nonsense," multi-floor **W Oparach**

Absurdu is most memorable for its madcap clutter: vivid murals, lampshades, furniture that a student would probably reject, tatty Persian rugs, and weird sculptures. There's a separate smoking room and a vague air of anarchy. But there are also Polish beers on draught (half-pints are forbidden) and live jazz, Balkan, blues, or swing of an evening. If craft beers are your thing, try **Jabeerwocky**, where 17 taps pour the best Polish brews.

Ząbkowska 6 (+48 66 078 0319, www.facebook.com/woparachabsurdu); Nowogrodzka 12 (+48 22 254 31 07, http://taproom.pl/)

☕ CAFFEINE KICKS

A million cultural light years away from the grandma pubs are Warsaw's growing canon of sharp brew bars. One of the coziest is *Relaks* in the leafy, southerly Mokotow district. While beans from domestic supremos (Kofi Brand) and premier European roasters (The Barn) are manipulated by award-winning baristas, there's no stinting on coffee-shop comfort. The counter gives onto a lounge space with vintage armchairs, rows of framed movie posters, window seats, tiled floors, and wooden desks. It's the sort of place in which you instantly want to hang out for a few snoozy hours.

Puławska 48 (no phone, www.facebook.com/ kawiarniarelaks)

MARKET RESEARCH

Warsaw's original food hall is back in business. With its Art Nouveau façade and colorful, steel-lattice features restored, *Hala Koszyki* has returned from the dead. This is arguably a gentrification-led kind of rebirth, however: while some (artisan) produce stalls remain, the focus is now on gastro offerings, dining, and the young-ish, cool, and moneyed. You've got very decent bistros like Warszawski Sen and a host of snacky options. Further upstairs is a modern-art gallery, design studio, and a light installation by Australian artist Michael Candy, all adding to the palpable sense of a place to be.

Koszykowa 63 (+48 22 221 8186, www.koszyki.com)

▼ In Hala Koszyki establishments serve everything from gelato, Thai, and tapas to pasta, sushi, and Polish-inspired cuisine.

Prague

CZECH REPUBLIC

> European cities don't come classier than Prague, whose elegant River Vltava bridges lead to a glory of cathedrals, cobbles, glitter- and gold-crested domes, and to that fairytale castle. While much cuisine there remains stodgy—and the service sulky—bold re-imaginings and a nascent Nordic scene have livened things up a great deal in recent times.

🍲 REGIONAL COOKING

While goulash and schnitzel (called řízek there) are omnipresent across the Czech Republic, *vepřo-knedlo-zelo*—roast pork, dumplings, and sauerkraut—is more of a local mainstay. All three are on offer for bargain prices at the central *Lokál Dlouháááá*. Complete with basic, refectory-like seating, this bustling homage to Czech trad divides its protein-tastic menu into regions of the country, buffeted by lagers and wine. Skip dessert and walk (wobble) to *St Tropez*— a patisserie typically selling nougat, caramel, and vanilla confections. Then fall over.

Dlouhá 33 (+420 734 283 874, www.lokal-dlouha.ambi.cz); Vodičkova 30 (+420 603 333 338, www.cukrarnatropez.cz)

ON A BUDGET?

The city boasts a longstanding Vietnamese community, a scenario which has inevitably led to some superior South-east Asian eats. Particularly cheap are the three branches of *Pho Vietnam Tuan & Lan*, whose street food tends to cost 60–120 koruna. No wonder locals willingly line up for the punchy beef noodle soups and fried pork sticks. Back toward the Vltava, near Charles Bridge, is *Bistro 19*—part minimalist interiors store, part gallery, and part bargain lunch spot where inventive soups followed by one of the two mains—meat or veggie—is only 100 koruna.

Anglická 15 (+420 606 707 880, www.facebook.com/photuanlan); Karoliny Světlé 19 (+420 739 466 838, www.no19.cz)

SPLASH OUT

Divinis, *Alchymist*, and *Bellevue* may attract the celebs, but nowhere does more daring and one-off dining better than La *Degustation Boheme Bourgeoise*. Its

▲ Eska uses cooking methods such as fermentation, drying, wood-heating, and fire-roasting to prepare its dishes.

chef, Oldřich Sahajdák, has garnered a Michelin star for his takes on recipes from an esoteric 19th-century Czech cookbook. As his Old Town joint's name suggests, he offers only two tasting menus: one six courses, the other a mammoth 11. Their contents change with the seasons, but might include boar steeped in juniper berries or catfish with kefir, poppy seeds, and dill. Naturally, local pinots, muscadets, and sauvignons also feature.

Haštalská 18 (+420 222 311 234, www.ladegustation.cz)

HIP & HAPPENING

Nowhere in the eastern hipster hub of Karlín has engendered more debate and disagreement than *Eska*. Still, despite being part of the moneyed Ambiente group—

which also includes *La Degustation*, from where it nabbed a sous-chef—this airy, high-ceilinged ex-factory has now just about seduced the cool crowd. How? By offering something very un-Prague: vegetable-heavy, Nordic-style cuisine. Hence the roasted carrots beside parsley root and mustard; hence the fermented red wheat with mushrooms and egg. Meats remain available for carnivores and Czech traditionalists, alongside output from an onsite bakery and superb coffee.

Pernerova 49 (+420 731 140 884, www.eska.ambi.cz)

LOCAL SECRET

Though located in the Old Town, *U Benedikta* skulks down an almost-eponymous side street and is somehow near-exclusively populated by Czechs rather than hordes of tourists. The restaurant is a vintage tavern, which serves vintage fodder: pork knuckles (ham hocks), beef flanks, beef shanks, and so on. If you can't decide what to have, have everything, or at least most things, courtesy of a "Bohemian platter," which chucks in some stroganoff for good measure. Portions are huge and prices are moderate, including for the various superb Czech beers—not least the excellent unfiltered Bernard lager—and lethal pear brandy.

Benediktská 722/11 (+420 224 826 912, www.restauraceubenedikta.cz)

BRILL FOR BRUNCH

Misto is the third café in the group—one which also spans *Muj Salek Kavy* (see right)—overseen by local roaster Doubleshot. All are flipping popular and requiring of reservations; the food, however, is better at Misto than at those two siblings by dint of having a bigger kitchen. Sunday brunchers can order from a blessedly light, low-protein menu—eggs Florentine, granola, buckwheat pancakes, gruyere omelets. There's even a Full English. Located in residential lands near the Hradcanska subway stop, Misto also scores for its interiors—a pale-wood dream—and ace coffee.

Bubenečská 12 (+420 727 914 535, www.mistoprovas.cz)

CAFFEINE KICKS

If it's just coffee you're after, though, Karlín's *Muj Salek Kavy* is the place—and that despite its name translating as "My Cup of Tea." There you'll find all the standards: artisan (Doubleshot) beans, whizzy machines, bearded and tattooed baristas, Instagrammable mural walls, organic this, gluten-free that. The unusually smiley staff train at an adjacent school, one viewable from palm-shaded tables outside on the snoozy pavement. In another hip hood, riverside Letná, *The Farm* also serves Doubleshot in modish, utterly on-trend (think exposed wood, large pipes, monochromed menu) surrounds. Farm-to-fork brunches are offered, too.

Křižíkova 105 (+420 725 556 944, www.mujsalekkavy.cz); Korunovačni 17 (+420 773 626 177, www.facebook.com/farmletna)

QUINTESSENTIALLY PRAGUE

Prolific in Prague are rooftop terraces with glorious panoramas. Predictably, these are far from cheap perches to frequent—many of them crown five-star hotels, including the best, *Terasa U Prince*. Found above a hotel of the same name on Old Town Square, this fine-dining resto offers oh-bloody-hell vistas of

▼ The terrace at U Prince—nestled between the russet-colored roofs of the Old Town—provides the Prague's premiere rooftop dining space in a city filled with rooftop terraces.

the cutely chiming astronomical clock, a sea of terracotta rooftops, various River Vltava bridges, and Prague Castle. Higher still is the south-easterly Žižkov TV Tower, whose topmost observation deck and also-posh *Oblaca Restaurant* stand 217ft (66m) high. Suffice to say, the views are delicious.

Staroměstské 460/29 (+420 224 213 807, www.terasauprince.com); Mahlerovy Sady 1 (+420 210 320 086, www.towerpark.cz)

▲ Náplavka Farmers' Market is situated on the banks of the river Vltava.

BEER OH BEER

Practically every Prague pub and restaurant serves Pilsner, and most of them do 45 varieties. It was invented in the Czech Republic, after all. Craft beer, like everywhere else on Earth, is also de rigueur, and increasingly prominent. Enterprising establishment *Nota Bene* is both a gourmet restaurant and artisan ale bar. Six taps froth out creations by Czech microbrewers—Matuška beers are particularly worthy of your attention. The daily-changing menu, meanwhile, sources from pre-eminent farms and producers. Nowhere else in Prague excels quite so well at both disciplines.

Mikovcova 4 (+420 721 299 131, www.notabene-restaurant.cz)

MARKET RESEARCH

Organic farmers' markets are another feature of the city. Handily held on weekends, *Jiřák* (Wed–Fri, 8am–6pm; Sat, 8am–2pm) sets up opposite the Church of the Most Sacred Heart of Our Lord, an imitation of Noah's Ark by Slovenian architect Josip Plečnik, and a few blocks from sprawling Rieger Gardens. Local restaurants hawk global produce from Slovenian to Thai, with the kolaches and sausages especially worth checking out. Brill for breakfast

is riverside *Náplavka Farmers' Market* (Sat, 8am–2pm): BrewBar coffee, artisan yogurt, and dozens of donuts are sold, and provide ample excuse to sacrifice that lay-in.

Náměstí Jiřího z Poděbrad (+420 770 125 373, www.trhyjirak.cz); Náplavka (no phone, www.farmarsketrziste.cz/naplavka)

Zurich

SWITZERLAND

> **The Swiss capital's Disney-like castles, superior museums, and lovely lakefront are accompanied by excellent restaurants. What with so many financiers visiting and Switzerland's famously expensive franc, you can expect to spend a lot—although a lot of good, informal, mid-range options have emerged in recent years.**

ON A BUDGET?

Spending only ten francs on anything resembling a meal in Zurich is one of life's great challenges. One of the city's few low-cost oases is *Basilikum*, a no-frills counter café where students scoff sandwiches which are a) large, b) excellent, and c) ten francs exactly. You choose focaccia—perhaps tomato seed?—and a filling, with charcuterie, cheese, pâté, pickles, and marinated vegetables all on offer, plus about ten possible sauces. A next-door flower store is responsible for blooms which brighten up the interiors, and there are superb city views.

Haldenbachstrasse 2
(+41 44 261 3234,
www.basilikum.ch)

SPLASH OUT

There are plenty of Michelin-starred lakeside lovelies and fancy ex-warehouse diners to choose from, but none can compete with **Sonnenberg Restaurant** for views. From its perch on Sun Mountain, you'll be able to Instagram magnificent panos of snow-crested Alps, Lake Zurich, and the city's clustered rooftops. Also known as the "FIFA Restaurant"—it has official links to football/soccer's governing body, which has its headquarters close by—Sonnenberg proffers a Mediterranean-Swiss-inspired menu of lobster with green beans and grapefruit, and venison stew beside pizokel and red cabbage. Euro guide *Gault Et Millau* has consistently showered praise and points on the results.

Hitzigweg 15 (+41 44 266 9797, www.sonnenberg-zh.ch)

HIP & HAPPENING

While any guide that calls Zürich West, or Kreis 5, gritty or edgy is about five years behind the times, this 'burb remains the city's prime nightlife hub and a still-cool blend of converted garages and railroad sidings. It includes **Frau Gerolds Garten**, an al fresco space where canvas tents, long benches, and shipping containers are used to host a restaurant, bars, and general hang-out for the young and beautiful. Fairy lights, outdoor terraces, grilled sausages, and home-grown salads dominate in summer; winter sees blankets and piping pots of fondues.

Geroldstrasse 23/23a (+41 78 971 6764, www.fraugerold.ch/gastronomie)

LOCAL SECRET

Getting to **The Artisan** requires a semi-schlep north into residential Zurich, but the few tourists who venture this far are rewarded with a lovely garden setting.

Inside, bright handfuls of hanging flowers droop from every possible angle overhead. In keeping with the greenery, menus champion everything local and seasonal. Vegetarians will relish vegan burgers and quinoa pastas, while carnivores can chow down on gingery, charcoal-cooked poussin. More central, in the Werd quarter, is family-run Sri Lankan haunt **Neela**. Mild curries come accompanied by *puttu* (small pasta made from red rice-flour) and just-right roti.

Nordbrücke 4 (+41 44 501 3571, www.theartisan.ch); Zweierstrasse 55 (+41 44 242 7111, www.neela.ch)

▼ The open-air Frau Gerolds Garten repurposes old shipping containers into cool bars and places to eat.

🍴 BRILL FOR BRUNCH

Once a butcher's, **Kafischnaps** is now famous for vegan muffins. And brunch in general: served there until 4pm, it also stars chunky, open sandwiches of roasted veal, tomato-chili jelly (jam), arugula (rocket), and hard Sbrinz splitter cheese; *ofenei* (oven-baked fried egg); and baked *tomme vaudoise* and fig chutney. Make sure to try the titular house schnapps, made from home-grown quince, too. Located just north of center in Kreis 6, the bright, black-and-white-tiled room and its retro chandeliers fills quickly on weekends, so reserve tables if you plan to visit then.

Kornhausstrasse 57 (+41 43 538 8116, www.kafischnaps.ch)

🍲 REGIONAL COOKING

Zurich's flagship feed is *geschnetzelte*: pan-fried veal in a rich, creamy sauce of white wine and mushrooms. Try it with rosti potatoes In the lime-trapped **Zeughauskeller**, a 15th-century armory which retains some ancient military accessories. Foamy ales wash the stew down. To try other Swiss standards, pricey **Alpenrose**—a Heidi-style dream of wood paneling and ornamental ceilings—is the place. Its meatloaf hails straight from the country's cattle-happy meadows, as do *spätzli* (cheese-topped egg noodles) and buttery carrots, buffeted by a Europe-wine wine list.

Bahnhofstrasse 28a (+41 44 220 1515, www.zeughauskeller.ch); Fabrikstrasse 12 (+41 44 431 1166, www.restaurantalpenrose.ch)

📋 QUINTESSENTIALLY ZURICH

"Mmmm, cheese." If this is a phrase you regularly utter, get yourself to **Holy Cow**. If you like great chicken and beef burgers, sprint there, ideally to the best, Rathaus branch. This Old Town institution tops many of its 20+ organic, gourmet patties with a range of cheeses: Gorgonzola, blue, Cheddar, and Camembert. Vegetarian and non-dairy options are also available, while top-notch Swiss meats are used throughout. Prices are pretty good for Zurich, too—a Viva España (toasted goat's cheese, honey-glazed chorizo, and hot chili) costs CHF 14.90.

Zähringerstrasse 28 (+41 21 312 2404, www.holycow.ch/holy-cow-zuerich)

🍸 KILLER COCKTAILS

Good luck finding a classier bar than the **Kronenhalle**. Together with its adjacent restaurant, this Zurich mainstay in the Old Town boasts high-top wooden counters and glittering glass, and originals by Picasso and Chagall. There are masters behind the bar, too, making cocktails and house libations such as the Kronenhalle Royal: Grand Marnier, crème de cassis, Angostura, and champagne. For something slightly cheaper and much livelier, try **Barfussbar**: women's-only baths by day, hipster bar-deck at night. Tea-lights and single-storey, Art Nouveau stylings make for a magical, most unusual venue. Wednesday has live music performances, on Thursday the bar is open, and on Sundays

▼ *Geschnetzelte* served with a rosti as a side dish: at this giant size it could easily be considered the main course.

▲ Barfussbar operates a unique "no shoes" policy.

you will hear local DJs spinning an eclectic mix of tunes.

Rämistrasse 4 (+41 44 262 9911, www.kronenhalle.ch/bar); Stadthausquai 12 (+41 44 251 3331, www.barfussbar.ch)

 CAFFEINE KICKS
Thanks to just the right balance of industrial chic and snug homey-ness, **Bros Beans & Beats** is—despite its horrendous name—the kind of place you want to enjoy for hours. Plain concrete walls are softened with low-hanging lamps and plenty of timber. Sourced from local roaster Henauer, the almond cappuccinos and filter brews are equally on-point. Not that it's all about coffee: you can

come for beers of an evening, enjoy a hot Reuben (pastrami, melted cheese, and sauerkraut on toasted sourdough)—come at lunchtime for those—or yank cheesecake from the vitrine at any (weak) moment.

Gartenhofstrasse 24 (+41 44 543 6580, www.facebook.com/brosbeansbeats)

MARKET RESEARCH
There's an international flavor to **Helvetiaplatz Markt** (Tues and Fri only, 6–11am), held on the eponymous square in multicultural area Kreis 4, with everything from Greek feta to Indian kumquats likely to be on sale. Home-grown produce is in abundance too, including fish freshly hooked in Lake

Zurich. Altogether more pop-up is the **Bahnhofmarkt** (selected Wed, 10am–8pm): a most-weeks farmers' market that takes place between April and November in main railway station Zürich Hauptbahnhof's hall and extends into the plaza. Regional Swiss specialties dominate there: particularly cheeses and meats.

Helvetiaplatz (no phone, www.zuercher-maerkte.ch/helvetiaplatz.html); Bahnhofplatz 7 (no phone, www.sbb.ch/bahnhof-services)

Vienna

AUSTRIA

> Vienna's grand Habsburg palace and Mozart-playing concert halls are matched by elite restaurants and ornate, cake-tastic cafés. But you've also got homely taverns and their schnitzels, superb sushi, sausage stands, mad-scientist cocktails, and artisan donut-makers. In other words, the food scene's as compelling as everything else in Austria's one-off capital.

💰 ON A BUDGET?

A definite Asian focus awaits at the opposite end of the Vienna spending spectrum. *Natsu* now has three branches in a growing sushi empire, but the one near the 6th district's aquarium is nicest. Join students there at lunch, enjoying half-price norimaki sets, various bento boxes, soba noodles, udon or ramen bowls, tempura, and so on. Whether your chosen feed is discounted or not, prices are low, but quality distinctly high. Eight pieces of vegetarian sushi costs under €4, for instance. Takeaway is entirely possible for tourists in a hurry, including from *Natsu*'s new MuseumsQuartier kiosk.

Gumpendorfer Straße 45
(+43 1 581 2700,
www.natsu-sushi.at/japan-lokal-6)

SPLASH OUT

Have you heard of Heinz Reitbauer? You ought to have: he's one of the world's great chefs, with his *Steirereck* making the top 10 of 2017's "World's 50 Best Restaurants" list and garnering two Michelin stars. That's the reward for refined dishes which give a contemporary, often-Asian makeover to vintage Austrian staples and serve them looking as pretty as the nearby Kursalon music hall. Menus change frequently to ward off imitators, but expect confections like Danube salmon with gooseberry jus or poppy-seed noodles. Finish with some local cheese, and then walk it off in surrounding Stadtpark.

Am Heumarkt 2A/Stadtpark
(+43 1 713 3168,
www.steirereck.at)

HIP & HAPPENING

At *Labstelle*, a similar mentality to Steirereck's— essentially revising staple Styrian plates and ingredients—is married with oh-so-sharp design. You'll eat your food amid black steel, shiny concrete, and Hans Wegner's Y Chairs—or in a hanging garden if the weather allows. Less swanky, yet just as cool, is *Donuteria*, whose enterprising flavors range from lime and mint to marzipan and chocolate (a "Mozart"). You can even order apple strudel-flavored rings. Find these artisans in the Haus der Musik.

Lugeck 6 (+43 1 236 2122,
www.labstelle.at);
Seilerstätte 30/Top 2
(no phone, www.donuteria.com)

▼ Steirereck offers suitably impressive surroundings for a restaurant considered by those in the know to be one of the world's best.

LOCAL SECRET

As apps like Vizeat and EatWith demonstrate with their Airbnb-for-foodies concepts, private dining in local homes is all the rage—it's authentic, insidery, and, let's face it, just plain fascinating. With its six serviced tables and outside courtyard, *Hofzeile 27* sits somewhere between those supper-club options (a little daunting for some) and a proper restaurant which holds events. Here, guests can admire Sibylle Fellner-Kisler and family's villa and interior decoration showroom, before sitting down to enjoy the likes of braised pork roast, fine dumplings, and sour cabbage. The villa is located in Döbling, a posh and leafy suburb to the north of the city.

Hofzeile 27/1 (+43 664 527 7929, www.hofzeile27.at)

REGIONAL COOKING

Wiener (veal) schnitzel is Austria's classic dish, shortly followed by *tafelspitz* (prime boiled beef) and goulash. There are many classic, homely "beisls" in which to try these dishes, including *Pürstner* and affordable *Reinthaler's*. The best, though, is *Zur Eisernen Zeit*, near the Naschmarkt. Its beef goulash bursts with brothy flavor, and its schnitzels are perfectly tender. Chat with locals, order gnocchi sides, and wash everything down with a big stein of beer. A junkier alternative is provided by Vienna's *würstelstände* (sausage stands) and their cheese-filled *käserkrainer*. *Alles Walzer, alles Wurst* (open daily, 8pm–5am) is the best.

Naschmarkt 313–316 (+43 1 587 0331, www.zureisernenzeit.at); Quellenstraße 84 (+43 664 212 8289, no website)

▼ A trip to sample the best of Viennese food could not be complete without trying the famous Wiener schnitzel.

▲ Much like their German cousins to the north, Austrians are wild for wurst. In Vienna you are never far from a neon sign signaling sausage in various forms.

BRILL FOR BRUNCH

Fact: the best brunch menus are available all-day, negating any need to rush and the chances of queues. *Phil*'s offerings are only available until 4pm, but that's still an acceptably long window of opportunity. A wide variety of food is on offer, including meats, cheeses, croissants, marmalades, eggs, falafels, and mueslis. The retro café is also a junk store and contemporary bookshop; a reliably in-crowd are present at all times. If there's no space available, walk north to *Figar* for avocado on toast. The toast in question utilizes fresh bread from the wonderful Joseph Brot bakery.

Gumpendorfer Straße 10 (+43 1 581 0489, http://phil.business.site); Kirchengasse 18 (+43 1 890 9947, http://1070.figar.net)

✅ QUINTESSENTIALLY VIENNA

It's a cliché, yes, but for good reason. Coffee and cake is the best Viennese ritual of them all. First, choose a suitably grand, historic *kaffeehäuser* such as **Café Sacher**, marbly **Café Central**, or, my favorite, refined **Café Landtmann**—straight out of a Wes Anderson movie. Sit there, in front of the Rathaus, and you'll be following

▲ To experience the historic Café Landtmann is to truly get a sense of the unique Viennese café culture.

in the discerning behinds of Burt Lancaster, Paul McCartney, Sigmund Freud, and Marlene Dietrich. Next, order a *wiener melange* (espresso with milk) and either a chocolatey Sachertorte or apple strudel with vanilla sauce. Then, indulge.

Universitätsring 4 (+43 1 24 100 120, www.landtmann.at)

KILLER COCKTAILS

What do clams, apple yogurt tea, and pickles have in common? They're all occasional ingredients for the wacky, Wonka-like cocktail concoctions of Kan Zuo. Home-brewed gin also features, while the vessels might be baby bottles or imitation popcorn boxes. Lurking up the top of Liechtensteinstraße, in Vienna's sleepy 9th district, **The Sign** lounge is as daring as mixology bars come. But it's also welcoming and warm, and deeply popular—meaning you'll need to reserve on weekends. Much more conventional drinks, from caipirinhas to lavender martinis, are available if you don't fancy the mad scientist stuff.

Liechtensteinstraße 104–106 (+43 664 964 3276, www.thesignlounge.at)

CAFFEINE KICKS

Jonas Reindl has plenty going for it. Its owner, Philip Feyer, directly sources from and supports sustainable plantations in Nicaragua. The baristas are knowledgable, the technology excellent, with guest roasts and drip-filter coffees well worth their €4. The atmosphere is cozy, courtesy of vintage furniture, sofas, working university students, bookish lecturers,

and the odd tourist. There's no pretentiousness, not even if you request a "Vienna Coffee" (two espresso shots under whipped cream). And, perhaps best of all, it is open on Sundays.

Währinger Straße 2–4 (no phone, www.jonasreindl.at)

MARKET RESEARCH

Not to be confused with the adjacent, same-named Saturday flea market, the outdoor **Naschmarkt** (Mon–Fri, 6am–7.30pm; Sat, 6am–5pm) fuses Central European and Balkan influences, and thus feels like a genuine crossroads. It also feels utterly exotic and chaotic, in the way all great food markets do. There, Turkish butchers hawk lamb-sausage ribbons next to Polish candy

stores or Croatian jam-makers. Homebrew honey wines compete with apricot oils and stone chocolates. There are meal stands too, some selling *palatschinken* (Austrian crepes, both savory and sweet) and oodles of liquor stores.

Between Karlsplatz and Kettenbrückengasse (no phone, www.naschmarkt-vienna.com)

◀ A serving of Vienna Coffee with its distinctive pillow of whipped cream sat on top of a double shot of caffeine-rich espresso.

Ljubljana

SLOVENIA

> Wedged between Italy, Austria, and the Med, Slovenia is easily missed. But that would be a mistake, especially for gourmands: the country is home to our planet's foremost female chef, while capital Ljubljana has a burgeoning repertoire of excellent restaurants and coffee bars to go with its castles, Cubism, and cute red roofs.

ON A BUDGET?

Eight euros. That's how much (or how little) the three-course lunch menu costs at *Pri Škofu*, a bistro run entirely by women. For this economical pleasure, you'll need to venture into the sleepy Krakovo neighborhood south of the center, and be prepared to eat next to equally parsimonious students. Reserve ahead and ask for the sunny terrace. Dishes span home-cooked favorites such as pork tenderloin in fig sauce and salads sourced from nearby allotments, but the best course is the last one—be it rich chocolate mousses or cherry-and-cheese dumplings, which are startlingly, life-changingly flavorsome.

Rečna 8 (+386 1 426 4508, www.facebook.com/gostilnica. priskofu)

SPLASH OUT

Strelec's setting, inside a high turret of Ljubljana Castle above the city, and the decor—rust-colored stone walls, fur throws, wall tapestries of medieval battles—scream of yesteryear, and its cuisine initially follows suit. Highly rated young chef Igor Jagodic designed his menu with an ethnologist in a bid to represent Slovenia's history. Yet there's something entirely contemporary about the eventual results: buckwheat croquettes filled with duck liver, black walnuts, and fir tree foam or veal tongue beside horseradish mayonnaise. You can dine outside, admiring 360-degree views, then ride the funicular back downhill.

Grajska Planota 1 (+386 3 168 7648, www.kaval-group.si/strelec.asp)

REGIONAL COOKING

Named as 2017's Best Female Chef on "The World's 50 Best Restaurants" list, Ana Roš cooks from her family's inn and restaurant *Hiša Franko*, a two-hour drive from Ljubljana via the gorgeous Soca Valley. But don't despair if you've not got time for that; Roš and her sommelier partner co-own *Gostilna na Gradu* in the grounds of Ljubljana Castle, where one locavore tasting menu offers a "Walk across Slovenia." That includes porcini soup, sea bass fillet fonda with broccoli cream, roast lamb, and walnut biscuits with baked-apple ice cream. It's a bargain at €30.30.

Grajska Planota 1 (+386 8 205 1930, www.nagradu.si)

HIP & HAPPENING

A neo-bistro straight out of Paris, dinky *Monstera* sees Cordon Bleu-trained wizard Bine Volčič cook up weekly set lunches and seasonal seven-course tasting dinners. Not to mention paired Slovenian wines and home-brewed beer on draught. There are no à-la-carte options, so you just have to take the beautifully prepared grilled octopus and artichokes in red wine sauce, roasted beef cheeks, or black pudding with langoustines as they

triumphantly come. It happens amid smart, post-industrial interiors and a modishly open, no-waste kitchen—explaining why local trendies attend in well-dressed droves.

Gosposka 9 (+386 4 043 1123, www.monsterabistro.si)

LOCAL SECRET

Eating at *Taverna Tatjana* feels more like attending a dinner party at some exotic great aunt's house than frequenting a restaurant. Tables are arbitrarily arranged in the nooks of various vaulted rooms—plus a small rear courtyard—across a homely townhouse on the Old Town fringes. But boy can this

particular great aunt cook: fish is the specialty, as emphasized by black cuttlefish risotto and fabulously fresh mussels. For something non-Slovenian, try Sushimama, and its scorpion fish nigiri and wagyu beef carpaccio. Ana Roš calls it "the best sushi in town."

Gornji trg 38 (+386 1 421 0087, www.taverna-tatjana.si); Wolfova 12 (+386 40 702 070, www.sushimama.si)

▼ Taverna Tatjana's homely setting is charmingly chaotic, but the cooking is precise and unfailingly high in its quality.

▲ Ljubljana's food scene is relatively unknown to the outside world, which makes it the perfect destination for inquisitive foodies.

🚌❗ WORTH A TRIP

Just north of Ljubljana across pretty arable countryside awaits a foodie extravaganza which simply defies categorization. Forget lunches: "sessions" is a better way to describe the multi-hour, oh-so-hospitable afternoons on offer at *Gostilna Skaručna*. Together with his son, owner Slavko first drives around Slovenia in a beat-up car picking up meat and produce, then uses everything—apricots, asparagus, beef tongue—for trad-style banquets served at large convivial tables amid typewriters, vinyl, and heaps more antique clutter. The organic cviček (red wine from southwestern Slovenia) flows and strange music plays.

Skaručna 20, Vodice (+386 1 832 3080, www.skarucna.si)

✅ QUINTESSENTIALLY LJUBLJANA

Ana Roš might be the current cover girl of Slovenian gastronomy, but its godfather is Janez Bratovž—the country's first cook to truly score global acclaim. Bratovž's still going strong, and cooking up local game and fish in his formal *JB Restavracija*. Look out for the beet (beetroot) ice cream, the horseradish pannacotta, and, a city classic, cottage-cheese pancakes with tarragon. Another staple is, bizarrely, horse meat, which is popular across the country and supposedly very healthy. You can try equine burgers and rolls at the *Hot Horse* kiosk beside Park Tivoli.

Miklošičeva 19 (+386 1 430 7070, www.jb-slo.com); Celovška 25 (+386 31 709 716, www.hot-horse.si)

🍷 PERPETUAL WINE

Though only intermittently encountered overseas, Slovenia's wine is superb. The same goes for *Dvorni Bar's* location, right next to the picturesque River Ljubljanica and its leaf-shrouded stone sides. All the country's best tipples are available to try there, with over 90 by the glass—the oaky, dry white Rebula from near Italy; intense Teran out of the Karst region; a honeyed pinot grigio called Sivi. If anything impresses, there's a store next door with bottles to take home. Lots of warm and cold tapas keep stomachs safely lined.

Dvorni trg 2 (+386 1 251 1257, www.dvornibar.net)

☕ CAFFEINE KICKS

Tine Čokl is endlessly impressive. A self-taught roaster and barista, he runs both the co-operative Buna, which distributes his coffees to other local cafés plus educates students on fair-trade and environmental issues, and his own zero-waste *Cafe Čokl*. Located opposite Ljubljana's flower market, its wood-paneled room is low-lit and cozy. If you prefer your coffee bars more millennial in spirit, then follow the river south to *TOZD,* whose riverside space has the obligatory bicycle

positioned on the obligatory exposed-brick wall. Locally roasted Escobar coffees served in pink cups are supported by Ruster, their wonderful cold brew, and bowls of homemade soup.

Krekov trg 8
(+386 41 837 556,
www.cafecokl.si);
Gallusovo nabrežje 27
(+386 40 728 882,
www.tozd.eu)

▲ Odprta Kuhna, where you can enjoy Slovenian fare and street-food classics, plus local wines and craft beer.

MARKET RESEARCH

Held on Fridays between mid-March and October, popular *Odprta Kuhna* (Open Kitchen, 10am–9pm) sees over 30 food stalls pop up in charming Pogačarjev square, beside the existing Plečnik market and its classy colonnades. The temporary affair sees global cuisines, plus outposts of *Strelec* and other local restaurants, offering cheap portions of their various fare. Lots of beer and wine's buyable, too, and live music is common. If you can't make it,

the open-air *Glavna Tržnica* (Central Market— Mon–Sat, 6am–4pm; Sun, 9am–6pm) operates daily and sells plenty of possible presents, including honeys, jams, and so forth.

Pogačarjev trg 1 (no phone, www.odprtakuhna.si/en); Vodnikov Trg 6 (no phone, www.lpt.si).

Zagreb

CROATIA

> While Dubrovnik and the Dalmatian Coast lure beach-bums and *Game of Thrones'* fans, compact capital Zagreb is Croatia's best bet for gastronomes. Nikole Tesle street is pretty much all bistros, while a small army of wine bars, cool coffee shops, and hidden courtyard restaurants dot the charming, red-roofed remainder.

💲 ON A BUDGET?

How do you feel about cottage cheese? One of Zagreb's staples is *štrukli*: baked filo dough and heaps of cottage cheese, served either boiled (sweet) or gratinated (salty). The best place to try it is at devoted diner *La Štruk*, found on a stepped alley close to the main bar street, Radićeva. One slab, which you'll struggle to finish, costs just 30 kuna. As well as the two traditional versions, twists here include štrukl with pumpkin-pesto, truffle, or blueberry. Eat your chosen variety on *La Štruk*'s patio, shaded by handsomely dilapidated stone buildings and a large tree.

Skalinska 5 (+385 14837 701, no website)

🔔 SPLASH OUT

Just north of Zagreb's hilltop Upper Town, home to a presidential palace, slopes away the serene Tuškanac forest. Amid its paths and pine trees intrepid *bon vivants* will find *Dubravkin Put*, a fine-dining haunt beloved by wealthier locals. At the restaurant you can choose from meat and fish tasting menus, alongside a-la-carte options galore. The fish on offer there is

▲ Štrukli can be found all across Zagreb. The excellent La Štruk serves both sweet and savory versions.

LOCAL SECRET
As Roman gods go, the Lares and Penates are hardly the sexiest: they were mostly household deities who performed plebeian tasks. But such an understated pair are perfectly emblematic for *Lari & Penati* the restaurant, whose informal atmosphere and incognito location render it easily overlooked. "Just another Zagreb bistro," most tourists think, hurrying by. Pause, however, and you'll quickly realize their mistake. Be it muffins, sandwiches, soups, fish fillets, or panko-fried chicken cuts, everything here is seasonal and high-quality. Given the relatively small capacity and loyal fanbase, lunchtime visitors should enjoy the best seat-securing odds.

Petrinjska 42A (+385 14655 776, www.faccbook.com/laripenati)

particularly acclaimed; hope that skate wings, cooked in pernod and tomato, and served beside eggplant (aubergine) and carrots, is available. To work up an appetite, you could first hike: an easy, pleasant path leads to beautiful Cmrok meadow, and takes about 90 minutes there and back.

Dubravkin Put 2
(+385 14834 975,
www.dubravkin-put.com)

HIP & HAPPENING
Croatian, Latin American, and particularly Japanese cuisines mingle seamlessly at *XATO*, an innovative bistro in the Scandinavian spirit, which is located just off main Ban Jelačić square. The city's sole robata grill—a multi-tiered, multi-temperature Japanese charcoal concoction—figures prominently amid all-organic four, five, and seven-course menus, each accompanied by wines or champagne. Offerings change every few days, but expect exciting plates like sirloin with green mole and a mayonnaise of scrambled egg, soy sauce, black oat, and sesame oil, usually served atop beautiful turquoise crockery. Reserve ahead in order to secure yourself a table on the al fresco terrace.

Petrinjska 2
(+385 91509 3191,
www.facebook.com/xato.hr)

 BRILL FOR BRUNCH

Going out for breakfast or brunch is a relatively recent craze in Zagreb, but several places have embraced the trend with relish. Most are found along Nikole Tesle, the main restaurant hub, and that includes *Bistroteka*, whose tables—like every other joint in the city—spill outside. Most popular mid-morning are its poached eggs on sourdough toast beside spinach, bacon, and more cream cheese, or scrambled eggs mixed with smoked salmon and delectable wasabi cream. The health-conscious crowd go for bowls of millet flakes, chia seeds, apples, and cinnamon, while everyone slurps artisan coffee.

Nikole Tesle 14 (+385 14837 711, www.facebook.com/ restaurant.bistroteka)

REGIONAL COOKING

To reach the Old Town you can either plod up winding, leafy stairways or ride Zagreb's short, steep funicular. Near to its scenic summit, and indeed to the quirky Museum of Broken Relationships, is *Bistro Vještica*. Rather romantic, with its red lanterns and wood beams, this cosy place specializes in delivering stalwarts of central Croatia's

▲ Nikole Tesle at the top of his eponymous street, where you will find several of the recommendations on these pages.

hearty, meat-happy cuisine. Hence the schnapps-flambéed chicken, the bacon-wrapped pork medallions, the turkey breasts and plums, and the mushroom-and-buckwheat soup. Wash your chosen feast down with cans of Garden craft beer.

Vranyczanyeva 6 (+385 16267 509, www.vjestica.eu)

QUINTESSENTIALLY ZAGREB

"Dvorište" is a word you'll need in Zagreb: it means courtyards, and refers to the alley-fed rectangles and squares secluded inside every block. These once housed skilled laborers, but now cafés are increasingly the norm—places you either know about or miss. Inside one dvorišta, and

halfway up a staircase for good measure, is *Mali Bar*. Here, TV chef Ana Ugarković provides unusual, punchy fare—goat's cheese, radish, hazelnut and quail-egg salad, say, or smoked tuna prosciutto with saffron sauce. You can dine inside, but the canniest guests head up again to a balcony terrace with rooftop vistas.

Vlaška 63 (+385 15531 014, www.facebook.com/ malibarzagreb)

KILLER COCKTAILS

There are three pertinent reasons to undertake the 20-minute tram ride south (alight at Veslačka) to *A Most Unusual Garden*, and endure its sometimes slow service. Firstly, the plant-packed patio's two-tier treehouse, which fits around ten people. Second, encouraged by said structure, the happy Alice in Wonderland type of whimsy: think wiggly, chequerboard paths, random piles of books, wispy garden furniture, and festoons of fairy lights. And, last of all, a Hendrick's gin bar serving superb G&Ts. Flavors from cucumber to orange are available. DJs play jazz or house after sunset, as a mellow ambience takes hold.

Horvaćanska 3 (+385 91464 6900, www.facebook.com/amostunusualgingarden)

CAFFEINE KICKS

One of Nikole Tesle's courtyards, or dvorište, concludes with *Quahwa*, an ace Arabica roastery and café which typifies Zagreb's java obsession—locals here notoriously devote hours to the twin pursuits of coffee and catching up. It's rather dim inside, so sit and sip on the deck instead, and watch cats dozing in morning sun or residents watering ferns on their loft balconies. Speaking of greenery, another caffeine option awaits across town. Owned by a florist, *Velvet* accompanies its poppy-seed cakes and flat whites with pot plants and huge, vivid bouquets, plus silvery chandeliers.

Nikole Tesle 9 (+385 91253 4140, www.quahwa.net); Dežmanova 9 (+385 1 4846 743, www.velvet.hr)

MARKET RESEARCH

Named "Kumica Barica," a bronze at *Dolac Farmer's Market* pays tribute to the friendly peasant women who used to regularly walk for tens of miles in order to hawk their vegetables, fruit, and dairy produce here. Most drive nowadays, but are still present, along with sauerkraut sellers, butchers, bakers, and olive-oil vendors. All of which renders *Dolac* a bustling, sense-thrilling extravaganza, especially when chefs congregate at its two-floor hall and piazza each morning from 7am. Well-regarded *Pod Zidom* restaurant, a few steps away, constructs its daily menus based on *Dolac* purchases made that morning.

Dolac 9 (+385 16422 800, www.trznice-zg.hr)

▼ Dolac Farmer's Market sees all kinds of Croatian produce for sale under a sea of red parasols.

Bologna

ITALY

> Parma ham. Parmesan cheese. Balsamic vinegar. Tortellini. All foods that hail from the Italian region of Emilia Romagna—of which under-rated Bologna is the capital, and hub. You'll find them all amid this pretty place's 25 miles (40 km) of colonnaded porticoes and rust-red walls. No wonder they call it la grassa (the fat one).

Via Cartoleria 15 (+39 51 272 900, www.settetavoli.it)

💰 ON A BUDGET?
A longstanding osteria southeast of the city center, *Vâgh íñ Ufézzí* offers absurdly good prices given the quality. Altered daily, its succinct menu uses fresh ingredients to fashion simple dishes. You might encounter pork stew with onions and cauliflower, pumpkin tortillas, or a legendary chocolate-chip mascarpone. Three delectable courses, half a liter of wine, and coffee will cost you under €20. Told you: absurd. An alternative is pizza, which in Italy is usually a) affordable and b) terrific. Bologna's best is at *Nicola's*, situated on a snoozy, beautiful square just to the north.

Via de' Coltelli 9 (+39 51 296 1446, www.vaghinufezzi.it); Piazza San Martino 9 (+39 51 232 502, www.ristopizzanicolas.com)

💬 LOCAL SECRET
One thing you won't find is spaghetti bolognese: the real thing is tagliatelle con ragu, which is far nicer and richer, thanks to its meaty sauce being prepared in butter. Along with tortellini in brodo (broth), this dish is ubiquitous around the city, but never better than at *Sette Tavoli*, near Bologna's university—the world's oldest, don't you know. A quaint time capsule of a place, with yellow walls and wicker chairs, the restaurant also serves smoked salmon with potatoes and pork cheek and asparagus risotto. As its name suggests, there are just seven intimate tables.

SPLASH OUT

Bologna has lots of white tablecloth and fine-dining but, surprisingly, just one Michelin star. That belongs to the *I Portici hotel*'s eponymous haunt, where young Campanian maestro Agostino Iacobucci remixes local staples with zestier southern Italian and Mediterranean approaches. Hence treats such as pigeon breast with plum, quinoa, and kefir or red tuna, vegetables, and lemon aïoli across tasting and à-la-carte menus. Equally impressive is the former auditorium room, scented by oil candles. Reserving is vital; *I Portici* only opens at night from Tuesday to Saturday, accepting just 40 covers per evening.

Via dell'Indipendenza 69 (+39 51 421 8562, www.iporticihotel.com/ i-portici-restaurant)

HIP & HAPPENING

Where parts of the Santo Stefano basilica are some 17 centuries old, nearby gastrobar *Ruggine* is infinitely modern—even if its voguish, straight-outta-Brooklyn design ethos champions salvaged furniture and exposed stone walls dating back 200 years. In this former bicycle repair store, hidden along an alley, you'll also find large-portion salads of melon, lardons, and almonds,

▲ A vintage Italian box to store gelato wafers on display at Bologna's gelato museum.

plus pastas such as spaghetti with salmon and chicory. Creative types glug coffee in-between and there are cocktails once the light begins to die. A seductively chill atmosphere pervades at all times.

Alley Alemagna 2 (+39 51 412 5663, www.ruggine.bo.it)

ICE-CREAM DREAM

Sure, every Italian city says it makes the best ice cream—but how many have both a gelato museum and university? Precisely. Bologna's most famous places to buy the cold stuff are *Cremeria Funivia*'s two stores—but fame equals long lines, so head elsewhere. *Cremeria Santo Stefano* has a nostalgic '50s

vibe, intense sugary smell, and interesting flavors such as crema di limone (lemon custard) or cinnamon, amid the classics. *Il Gelatauro*, meanwhile, is more artisanal and affords seating; it also serves cioccolato all'arancia (chocolate orange) flavor, which surely makes it a winner in everyone's book?

Via Santo Stefano 70 (+39 51 227 045, www.facebook.com/ cremeriasantostefano); Via San Vitale 98B (+39 51 230 049, www.ilgelatauro.wordpress. com)

🍜 FOREIGN FODDER

The one criticism of Bologna's gastronomy is that, while delicious, it can be a bit repetitively rich. To break things up, indulge in the city's best Asian eats. Fusing Chinese and Japanese cooking in a north-of-center, well-worth-it location, *Ristorante Megu* promises superb sushi, Gyoza dumplings, pad thai, katsu curries, and nigiri rolls in a small, informal room with pagoda stylings. The best news? An "all-you-can-eat" option is available for €15, or just €10 at lunch. You can probably skip dinner altogether.

Via Della Grazia 35 (+39 51 991 6521, www.megu.it)

🪧 DELECTABLE DAY TRIP

It's found a little outside Bologna, but there's no ignoring *FICO Eataly World*. Best described as a large "Disneyland for foodies," created by the super-successful deli chain *Eataly*—big in Italy and the USA. Opened in November 2017, this is essentially a 100,000-square-meter foodie theme park. (The space is so vast bikes are provided to help visitors get around.) The stated aim is to provide a "fork to table experience," enjoyably showcasing the journeys

▲ With 25 acres dedicated to food, if you can't find what you're looking for at Eataly World, it's not worth having.

made by various foodstuffs. Highlights include chocolate fountains, authentic oil mills, a truffle area, workshops galore, a brewery, and some 20 restaurants. Bus 35 operates direct from Bologna's station, and entry is free.

Via Paolo Canali 8 (no phone, www.eatalyworld.it)

🍸 KILLER COCKTAILS

As with all of Italy, aperitivo culture is staunchly observed in Bologna, with pre-dinner snacks commonly combined with spritzes, cocktails, lagers, or local wines. For the latter, try *Camera A Sud*, whose piazza-side terrace is dreamily perfect for warm evenings. Lambrusco—forget the rowdy-teenager associations—and Sangiovese are the regional tipples, while craft beers are also served. If you'd prefer cocktails, *Pastis* is the place. Most of its award-winning drinks incorporate the titular anise liqueur. Look out for cutesy Moroccan tiles and a youthful, late-night crowd on weekends.

Via Valdonica 5 (+39 51 095 1448, www.cameraasud.net); Via Belvedere 7/C (+39 51 230 650, www.facebook.com/pastis.belvedere)

 CAFFEINE KICKS
It might look like a quaint English tea room, thanks to small lamps and framed newspaper prints, and it might even serve a range of *tè*, but *Caffè Terzi* is decidedly a coffee bar. Along with traditional Italian-style coffee, flat whites and the like are made with pure Arabica beans and guesting small-batch products. Then come the specialty drinks. Caffè con Cioccolato melts high-quality semi-sweet (dark) chocolate shavings atop hot espresso; Caffè alla Nocciola adds hazelnut cream to the same pairing; and Caffe Arancia e Amaretto blends dark roast with orange and almond liqueur.

Via Guglielmo Oberdan 10 (+39 51 034 4819, www.caffeterzibologna.com)

MARKET RESEARCH
Like many global food halls, *Mercato di Mezzo* has seen a recent renovation and subsequent reinvention into a plusher, more touristy-friendly existence. Very close to Piazza Maggiore, Bologna's focal point, it was the city's first ever indoor market, but had been largely abandoned in the early part of this century. Today, however, the three-storey pavilion is thriving. Small stalls serve everything, from sausages to sorbet; and there's an artisanal beer pub in the basement and a first-floor pizzeria. Products remain on sale from rather chichi enotecas and bakeries, and the busy place is open daily from 8:30am.

Via Clavature 12 (+39 51 228 782, no website)

▼ Given the wealth of produce made in and around Bologna, the city's "La Grassa" nickname is justified.

Naples

ITALY

> Most visitors fly into Naples and then journey on, bound for Pompeii or the Amalfi Coast. Bad move. Behind the brashness and brouhaha awaits a much under-rated food scene. Pizza was invented here, bejesus, while those zesty, southern Italian ingredients are ubiquitous. So why not stay a few days, and treat your tummy?

 QUINTESSENTIALLY NAPLES

There's only one place to start. Queen Margherita's visit in 1889 supposedly led to *Pizzeria Brandi*'s chefs naming a brand-new, tomato-and-mozzarella combination after her, and Italian gastronomy changed forever. Pilgrims can still visit Brandi and enjoy the same dish, while those in favor of gourmet revisions have two headline options. While *La Notizia*'s 60-second efforts and sausagey calzone are located in hillside neighborhood Vomero, *50 Kalò* has an outpost near the Bay of Naples. It uses impeccable local fare—*fior di latte* from Agerola;

pork from Caserta—to create delicate, caramelized affairs.

Salita Sant'Anna di Palazzo 1–2 (+39 081 416 928, no website); Piazza Sannazaro 201/B (+39 081 192 04 667, www.50kalò.it)

LOCAL SECRET

Inside the ritzy Chiaia area, you'll find luxury boutiques and leafy parks—and also, incongruously, *Da Tonino*, a family-owned backstreet osteria hiding behind wispy curtains. This is a place about which only natives know; the kind of time-trapped haunt where white-haired men wear cardigans, do crosswords, and passionately put the world to rights, and where the owner clinks glasses of grappa with patrons. Become one yourself by ordering *cicinelli* (or *bianchetti*)—teeny whitebait splashed with lemon and oil—or rigatoni steeped in ragù and ricotta from the handwritten menus.

Via Santa Teresa a Chiaia 47 (+39 081 421 533, no website)

REGIONAL COOKING

There are plenty of other Neapolitan dishes, from *friarielli*, a broccoli-like cabbage served with salami, to *polpette*, fried veal and-pork meatballs in slow-cooked ragù using tomatoes from nearby San Marzano. Have the latter in *La Cantina di Via Sapienza*, a lunch-only, cash-only osteria which offers affordable home cooking deep inside the centro storico. Be sure also to try two desserts: rum-soaked babà cakes and the crunchy, shell-shaped *sfogliatelle* pastries, piped with sweet ricotta and almond paste, and the sole offering at *Pintauro*—a small, cutely dilapidated *pasticceria* in the working-class Spanish Quarter—for over 225 years.

Via della Sapienza 40–41 (+39 081 459 078, www.cantinadiviasapienza.it); Via Toledo 275 (+39 081 417 339, no website)

▼ Polpette, or meatballs, cooked in a sauce made from San Marzano tomatoes. Grown locally, these are widely considered to be the best canned tomatoes in the world.

 MARKET RESEARCH

While natty Nonna slippers, Neapolitan hip-hop, colanders, and coasters are all available at the blaring *La Pignasecca* (open daily, 8am–1pm) street market, food is its chief focus. Stallholders loudly hawk every conceivable produce: still-wriggling squids here, tumescent zucchini (courgettes) there, 35 kinds of cous-cous farther on. Beyond plain gawking, tourists can present-shop at deli stands, with olive oils a choice purchase, or remove to one of the surrounding street-food stands. Best are the *friggitorie*, whose deep-fried delights include ham, cheese, or eggplant- (aubergine-) flavored *crocchè* (croquettes)—served in paper cones.

Via Pignasecca (no phone, no website)

▲ Those from colder climates can only dream of having easy access to the sort of fresh produce hawked at La Pignasecca.

$ ON A BUDGET?

With Naples being a harbor town at heart, seafood is another specialty. Situated next to La Pignasecca, **Pescheria Azzurra** is a fishmongers and restaurant by day, and solely the latter after dark; its original incarnation ensures the ingredients are super-fresh. Service can be on the slow side, but it's worth persevering, especially for anyone ordering the garlicky mussels to start and a subsequent *spaghelli alla vongole* (with clams), both of which are *eccellente*. Throw in bottles of Peroni and it'll still cost under €20 a head, including the tourist tax — not bad anywhere, and sensational somewhere this good.

Via Portamedina 5
(+39 081 1925 0592, www.
isaporidellapescheriaazzurra.it)

▼ With clams fresh from the Med and pasta cooked to al-dente perfection, spaghetti alla vongole is a must try.

SPLASH OUT

Fine-dining in Naples is also about flamboyance, and few chefs rival Lino Scarallo in that department. Like many haute-cuisine hubs, his *Palazzo Petrucci* gives traditional local fare an elaborate, decidedly modern makeover. Witness the linguine with smoked eel, licorice, and salted orange. Scarallo's flagship restaurant formerly occupied a 16th-century stables in the historical center's focal piazza, but that's now his highfalutin' pizzeria—famed for its remarkable "Bloody Pizza" drink (www.bloodypizza. it). The real thing has moved a few miles out of town to an ancient building romantically positioned beside Villa Donn'Anna beach.

Pizzeria: Piazza San Domenico Maggiore 5–7 (+39 081 551 2460, www. palazzopetruccipizzeria.it); Restaurant: Via Posillipo 16 (+39 081 575 7538, www.palazzopetrucciristorante. it)

ICE-CREAM DREAM

Ice cream's important anywhere, but in a city as full-throttle, chaotic, and—mostly—hot as Naples, it's especially so. There's great gelato all over, but good luck finding anywhere better than

Officina Gelati. The vast array of flavors extends to apple, frozen yogurt, and toast-with-Nutella, but two classics rule the roost: *cioccolato*, obviously, topped with wafers and nuts, and the very-Italian *stracciatella*, an extra-creamy, chocolate-chip concoction. Milkshakes and ice-cream brioche are also available, as is esoteric Grandpa Paul—blending egg yolk, sweet Sorrento lemon peel, cinnamon, and vanilla.

Via Toledo 311 (+39 081 405 312, no website)

▲ Rich, irresistibly creamy, and oh-so-smooth; Italians really know how to do chocolate ice cream.

FOREIGN FODDER

"Best sushi ever," according to more than one TripAdvisor comment about *Honzen*, a minimally decorated, 40-seat Japanese joint located near Villa Donn'Anna in Posillipo. The expansive menu—nigiri and sashimi through to ramen—majors in unusual creations, such as flambéed cobia fish or uramaki lobster with truffle pearls, and is accompanied by an extensive Italian wine list. A ten-course menu stands by for anyone unsure of what to order, concluding with some unexpectedly yummy cream tarts. Before you vroom off, cross the street and admire some wonderful views of the Bay of Naples.

Via Alessandro Manzoni 126 (+39 081 714 7201, www.honzen.it)

CAFFEINE KICKS

No coffees are posher than the ones served at *Gran Caffè Gambrinus*, on the stately Piazza del Plebiscito. Past regulars include poet Gabriele d'Annunzio and philosopher Benedetto Croce, while tradition sees the current President of the Italian Republic sip espresso there every New Year's Day. Far less ostentatious are the various branches of *Caffè Mexico*

dotted across the city, and especially its Piazza Dante outpost. An orange-colored machine serves sweetened blends to policemen, politicos, and punks alike, with orders of *harem con panna* (Arabica espresso topped with thick cream) most likely to ingratiate you with the chirpy baristas.

Via Chiaia 1–2 (+39 081 417 582, www.grancaffegambrinus. com); Piazza Dante 86 (+39 081 549 9330, no website)

KILLER COCKTAILS

The city's premier aperitivo spot is *Ba-bar*, a French vibe, café-bistro hybrid which comes alive at night. Soft jazz plays and the basement kitchen churns out salad and burgers, but you've come for libations: Moscow Mules (vodka, ginger ale, lime) is the house drink. Others are content to play backgammon, or people-watch. The one downside is the relatively high prices—this is Chiaia after all—but there are no such drawbacks in *Cammarota Spritz*, a student-popular, Spanish Quarter dive bar. Every plastic-cupped drink there costs but €1, even an Aperol spritz. With such a small space, the crowd spills outside onto a rowdy street.

Via Bisignano 20 (+39 075 456 812 10, www.ba-bar.it); Vico Lungo Teatro Nuovo 31 (+39 320 277 5687, www.cammarotaspritz.com)

▼ Apero hour is serious business in Naples. Make like the locals and order an Aperol spritz or a Negroni.

Palermo

ITALY

> Palermo's food is as maverick, impassioned, and bewildering as everything else in Sicily's madcap, crumbling capital. There are numerous signature dishes, from arancini to seafood spaghetti, and sprawling food markets; there are fried-food stands wafting scents with abandon and Arab-influenced pizzerias. There's even a drink called "poison."

⌐ REGIONAL COOKING

Nearly every restaurant claims to serve some take on *cucina alla Sicilia*, but two do it better than the rest. Palermo is a harbor town at heart, meaning oh-so-fresh seafood: the owners of unpretentious **Trattoria Piccolo Napoli** have two fishing boats themselves, and their daily catch decides offerings such as casarecce pasta with swordfish and mint or macaroni in sardines, wild fennel, pine nuts, and some apricot jelly (jam). Over at busy **Bisso Bistrot**, atop the Quattro Canti, the traditional antipasti on offer includes *panelle* (chickpea fritters) and *caponata* (spunky eggplant/aubergine salad), plus pumpkin ravioli as a main.

Piazzetta Mulino a Vento 4
(+39 091 320 431,
www.trattoriapiccolonapoli.it);
Via Maqueda 172a (+39 328
131 4595, no website)

⌐ SPLASH OUT

Shelter from Palermo's marvelous hurly-burly is on offer at **Gagini Social Restaurant**, whose name refers to its romantic location: the former workshop of Renaissance sculptor Antonio Gagini. The title also alludes to how diners are encouraged to mingle on shared tables, all as

a seasonal menu delivers modern European riffs on staple Sicilian fodder: saffron risotto and quail-and-carrot flan with spinaci "candy," say, or beef tongues beside mustard ice cream. By fine-dining there, you'll also be affirming the island's Slow Food Presidia involvement, which valiantly promotes regional farmers and producers.

Via dei Cassari 35
(+39 091 589 918,
www.gaginirestaurant.com)

▲ The stone walls in the candlelit Gagini Social Restaurant date back to the 15th century.

HIP & HAPPENING

You don't find many bar/restaurant/liquor store/bookstore/gallery/concert hall/club/Internet café/travel agency mash-ups around the world, and still fewer hidden inside ruined 19th-century palazzos built into seawalls. No surprise, then, that the cool kids flock to *Kursaal Kalhesa*. Its main spaces are a rooftop bar and lush courtyard where Aperol spritzes can be enjoyed, plus a grand, vaulted-ceilinged space in which contemporary

▲ Whether you go for the restaurant, the bar, or to dance, Kursaal Kalhesa offers unique experiences.

Sicilian-Arabic food is devoured. Order grouper-stuffed ravioli if it's available. The *Kalhesa* stays open until the early hours on animated weekend nights, but is wonderfully chilled otherwise.

Foro Umberto I 21a
(+39 340 157 3493,
www.kursaalkalhesa.it)

QUINTESSENTIALLY PALERMO

Friggitorie (fried-food stores) are a common sight around Palermo and typically hawk ragu-stuffed arancini, *pasticcino* (sweet pastry filled with mince), and *sfinciuni*, a soft flatbread topped with tomatoes, onions, anchovies, breadcrumbs, oregano, and sometimes cheese. These longstanding institutions offered fast food way before McDonald's arrived, and now burst with street-food credibility. Tiny *I Cuochini* is considered the grandmaster: wedged between two theaters in the old town, it serves the classics, plus more daring options such as *pani ca meusa*—cheesy rolls and, eek, sautéed tripe.

Via Ruggero Settimo 68 (+39 091 581 158, www.icuochini.com).

▲ Gelato served in a brioche bun—possibly one of the most ornate takes on the ice-cream sandwich you will come across.

ICE-CREAM DREAM

Yes, *Gelateria Ilardo Giovanni* does superb ice creams. In fact, it probably does Palermo's best gelato, once scoffed by Garibaldi himself. And yes, the old-school bar boasts a terrific location: at the base of the Cattive seawall, rubbled during World War II air raids, but now restored. Yes, yes, yes. But locals don't go there because of any of that, at least not solely; they chiefly go for the Ilardo's ice cream-stuffed brioche buns. Let's say that again, together: ice cream-stuffed brioche buns. Mulberry and whipped-cream flavor is the most popular, shortly followed by citron, strawberry, and pistachio.

Foro Italico Umberto I 1 (+39 091 617 2118, no website)

☕ CAFFEINE KICKS

Everything in Palermo tends to be full-on, and that includes coffee strong enough to wake the dead—although, in containing more arabica than usual, it's also less caffeinated. No caff is more celebrated than the award-winning *Antico Caffè Spinnato*, a place to which one can come all day. From 7am, locals chat and chug cappuccinos; later, espressos, macchiatos, and correttos dominate as the evening takes hold, before grappa is sipped until 2am. You can also purchase pastries whose cream uses nuts from Bronte, an eastern town which supposedly grows the world's finest pistachios.

Via Principe di Belmonte 111 (+39 091 749 5104, www.spinnato.it)

◀ A congregation of saints keep a watchful eye over the *melanzane* for sale at the Mercato di Ballarò.

and on offer at the dive bar-like ***Taverna Azzurra***, near to La Vucciria. For something rather more 21st century, head north to ***Tribeca***. Though the city's coolest aperitivo spot, this place is predominantly a sushi and sashimi restaurant, meaning cocktails have a Japanese bent. Expect sake and whiskey-infused blends amid artisan lampshades and atmospheric low lighting.

Via Maccherronai 15 (+39 091 304 107, no website); Via Mariano Stabile 134 (+39 091 332 963, no website)

MARKET RESEARCH

The city possesses two fantastic food markets. Most glorious is the ***La Vucciria***, an outdoor centro-storico behemoth where glistening fish line trestle tables, vast crates of artichokes sit on top of one another, tangs from tubs of harissa compete with porcini pongs, tomatoes are haggled over by unshaven chefs, and lines of salami hang off nooses. It's busiest at around 7am. The ***Mercato di Ballarò***, meanwhile, is more of a street bazaar, and especially good for cheese and olives. You can also buy ready-to-eat *babbalucci* (snails) in garlic, sucking them from their shells.

Piazza Caracciolo (no phone, no website); Via Ballaro 1 (no phone, no website)

KILLER COCKTAILS

The drink has two names, neither sounding promising: *veleno* translates as "poison," while *sangue* means "blood." A vermouth-based, vaguely Negroni-like shot best drunk over ice in combo with grilled meats, it's a Palermitan classic

PERPETUAL WINE

A rather urbane wine bar, cozy ***Enoteca BuoniVini*** has shelves densely packed with bottles from across Sicily, including vintages by the esteemed Planeta vineyard. Many are available by the glass, and best consumed alongside cheese and charcuterie boards. The inventive tapas also extend to foie gras "fillets" and, cutely, fish and chips. You can sit inside, on high tables or at the counter, or amid the terrace's colorfully cushioned seats fashioned from wooden crates.

Wherever you end up in the enoteca, a nice buzzy atmosphere permeates throughout. Be warned that service can be slow, and occasionally grumpy, but the drinks are well worth it.

Via Dante Alighieri 8
(+39 091 784 7054,
www.enotecabuonivini.it)

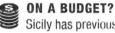 **ON A BUDGET?**
Sicily has previously been under Moorish and Byzantine occupation, and both influences still pervade at the marvelous *Mounir*, christened after its perma-smiling owner. This very reputable joint is a pizzeria and kebab house in one; it also sells large portions of couscous and vegetables. If you can't decide between the two main offerings, have 'em both in the curious—and very popular, in a cultish way—form of a kebab pizza, served with a side of Arabic mint sauce. There are a few seats outside or you can take away, and everything costs in the €5 region.

Via Giovanni da Procida 19
(+39 389 557 1525,
no website)

▼ The Taverna Azzurra offers an authentic Italian dive bar experience no hipster pretentions, it's the real deal.

Thessaloniki

GREECE

> You can keep your Athens. While it might be Greece's second city by size, Thessaloniki comes in first for gastronomy. There you'll find a dynamic scene, full of the verve of a university city, the cosmopolitanism of a Balkan gateway, the innovation of post-financial crisis survivors, and the sunny goodness of a Mediterranean climate.

💰 ON A BUDGET?

Everywhere in Thessaloniki is dirt-cheap, but if you're really keen to pinch the pennies, head for *Derlicatessen*—the name is a pun on a Greek word for eating. There you'll join a line waiting for huge pitta sandwiches containing bespoke blends of halloumi, mushroom, pork, chicken, feta, and salad, which barely cost €3 and taste fabulous. There's a smattering of outside seating. Similarly economical is *Pizza Poselli* in bar-lined Valaoritou, where slices of wood-burned, wafer-thin pizza are also €3. Look out for two quirks: chocolate calzone and the building's built-in clock, which stopped after 1978's earthquake.

Ioanni Kouskoura 7 (+30 231 022 6367, www.facebook.com/nterlikatesenolasoublaki); Vilara 2 (+30 231 401 9687, www.facebook.com/poselli)

🛎 SPLASH OUT

If anywhere can be labeled expensive in Thessaloniki, it's the attractive waterfront area. Primely positioned beside handsome and focal Aristotelous Square, *Agioli* is an award-hoovering brasserie whose menu offers a variety of Hellenic flavors and seasonal treats. Seafood is the

▲ The Ano Poli district, just outside the city center, gives you a sense of the real Thessaloniki and plenty of lunch options.

specialty: you might eat pan-fried red mullet, octopus carpaccio, smoked mackerel and lemon gel, or *midopilafo* (mussel risotto). Couples should try to snag a table on the romantic first floor, where views extend over the Gulf of Salonika to Mount Olympus, and bouzouki players sometimes strum their stuff.

Leoforos Nikis 15 (+30 231 026 2888, www.agioli.gr)

🗨 LOCAL SECRET
Charmingly scruffy, Ano Poli (Upper Town) is a jumble of wooden houses and cats snoozing under fountains. Fewer tourists walk uphill there, and thus bucolic tavernas like *Igglis* (a bastardization of "English") remain the preserve of boisterous locals. Traditional

Greek cuisine—more wispy pizza, slow-cooked pork shank, feta-stuffed roasted peppers, fava bean dips—along with apparently limitless grappa (*tsipouro*). Alternatively, take a taxi out to *Duck*, and enjoy the very Thessalonian contradiction of gourmet farmhouse-style cooking amid an industrial estate. Every meal should finish with their mandarin millefeuilles.

Irodotou 32 (+30 231 301 1967, no website); Halkis 3, Patriarchal Pylaias (+30 231 551 9333, www.facebook.com/duckprivatecheffing)

◣ SWEET TOOTH
Those Thessalonians sure do like a sugary snack. The Northern Greek mainstay is bougatsa, hot flaky pastry commonly filled with sweet cream or custard, but sometimes Merenda (Greece's Nutella), or savory with spinach, yogurt, mincemeat, and even Tabasco. It is typically accompanied by cartons of chocolate milk. For a fine, hand-crafted version, venture north to *Bantis* (Panagias Faneromenis 33); otherwise, head for all-night *Bougatsa Giannis*—a legendary, longstanding establishment. Many other cakes are available near the seafront in *Sugar Angel*: the lemon pies and raspberry cheesecakes are especially delectable.

Mitropoleos 106 (+30 231 025 7375, www.bougatsagiannis.com); Lassani 1 (+30 231 022 5575, www.sugarangel.gr)

🍲 REGIONAL COOKING

The bounty of Greece's islands is utilized at *Sempriko*, in Valaoritou, a co-operative restaurant and grocery store combo built into the original city walls. Mezze is the thing there, spanning classic lamb "manti" dumplings with yogurt, and newfangled plates such as smoked eggplant (aubergine) salad with tahini. Plenty of cheese and craft beers supplement the main dishes, and can also be purchased next door. Back in Ano Poli, peasant-style *Toixo Toixo*'s focus is the mainland, with pasta from mountain villages, and Macedonian wines served alongside chargrilled liver slices. Equally stupendous are the city views.

Fragkon 2 (+30 231 055 7513, no website); Polydorou Stergiou 1 (+30 231 024 5351, www.facebook.com/toixosalonica)

✅ QUINTESSENTIALLY THESSALONIKI

What Athenians—and most of us—call souvlaki or simply kebab, Thessalonians call a *gyro* (pronounced "yee-lo"). And gyros are ubiquitous in the city. Some of the tenderest pork or chicken, freshest tomatoes, tangiest onion, supplest tzatziki, and puffiest pitta is served up inside

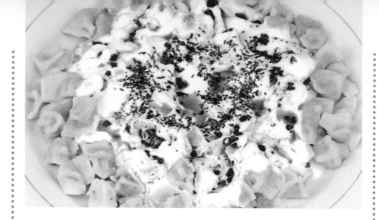

▲ Manti dumplings, served here with yogurt and chili oil, are popular everywhere along the Aegean coastline.

cave-like *Alaniara Kota*. Another staple is feta, and nowhere honors the cheese more than deli-restaurant *Mia Feta*. The "feta bites" there incorporate prosciutto, black truffle, and much more. Other Greek cheeses and lots of local wines anchor the offering.

Agiou Dimitriou 139 (+30 231 020 6010, www.facebook.com/alaniarakotathessaloniki); Pavlou Mela 14 (+30 231 022 1120, www.facebook.com/miafetafetabar)

🍸 KILLER COCKTAILS

Thessaloniki supposedly has more bars per capita than anywhere in Europe, and that certainly seems credible while walking around Valaoritou on any given night. For straight-up cocktail quality, *Vogatsikou 3* takes some beating: calling on a large apothecary of

bottles, its barmen craft homemade potions like the Pocket Rocket: vodka, passionfruit purée, coconut syrup, lemon juice, and, yes, arugula (rocket). Rooftop bars are common—try trendy *Urania* (Paikou 4) for views and late-night DJs—while oenophiles are directed to *Local*. Among the many Greek wines found there are Kir Yiannis, which is beloved of the mayor.

Vogatsikou 3 (+30 231 022 2899, www.vogatsikou3.gr); Paleon Patron Germanou 17 (+30 231 022 3307, www.localthessaloniki.gr)

☕ CAFFEINE KICKS

Rituals don't come much more Greek than lazily drinking coffee in the afternoon. Accordingly, you won't have any difficulty finding a traditional café. Third-wave places are scarcer, however, so praise be for *The Blue Cup*. A cappuccino's throw from the sea, it sees skilled baristas offer the house java—Mokka—plus a couple of guest single-origin blends, which can be bought. If it's hot, order a frappe; the drink was invented in Thessaloniki. *Blue Cup* morphs into a bar come the evening, while small School of Coffee classes teach roasting or incredible latte art.

Salaminos 8 (+30 231 090 0666, www.thebluecup.gr)

🍴 BRILL FOR BRUNCH

Estrella's celebrity offering is the bougatsan, chef (and self-proclaimed "food anarchist") Dimitris Koparanis' mash-up of bougatsas and croissants. But this street art-speckled space, located near the White Tower, offers far more than that gimmicky snack. Available all day, its globally influenced brunch menu also proposes waffles with smoked salmon, feta-spliced omelets, avocado and egg bagels laced in wasabi mayo, cheese and cherry tomato salads, bacon and eggs, brioche buns, and numerous pancake possibilities. Be warned that the service can leave something to be desired. It's worth persisting, though.

Pavlou Mela 48 (+30 231 027 2045, www.estrella.gr)

🎩 HIP & HAPPENING

"Gourmet Greek with a Cretan twist"—so says none other than Vogue of *Sebrico*, before also excitedly citing "a modern-industrial setting." But there are two main reasons why this Ladadika/Valaoritou haunt is so good, and so, well, in vogue. First it is, given the quality, bonkers cheap: €15 will feed you amply. Secondly, the hearty food is prepared by a collective of amateur, destined-for-greatness chefs keen to prove their chops. Every last slow-roasted pork knuckle, onion-fried egg, grape-must steak, pastrami pie, or vine-wrapped sardine is stupendous.

Fragkon 2 (+30 231 055 7513, no website)

▼ Tsipouro, a pomace brandy, is everywhere in Thessaloniki, often replacing wine or coffee as a drink. Be warned, it's potent.

Istanbul

TURKEY

> An unofficial position as Europe and Asia's geographical and cultural border post, coupled with superb seafood, has betrothed Istanbul with thrillingly exotic cuisine. To illustrate the point, just think about the country's clump of typical offerings: kebabs, köfte, baklava, pastirma, pide, dolma, raki, Turkish delight, Turkish coffee…

⬤ ON A BUDGET?

According to Istanbul tradition, workers dine at *esnaf lokantası*—tradesmen's restaurants—which serve up uncomplicated home-cooked fare at piffling prices. The waterfront Karaköy neighborhood may have been much gentrified in recent years, but **Nato** has barely altered its formula since opening in 1952. The menu changes daily, depending on what's available and what the chef fancies making, but expect bargain vegetarian stews, imambayıldı (stuffed eggplant/aubergine), pilafs, doners, and kuzu haşlama (braised lamb in light, lemony sauce). You'll probably have to share tables with students or actual, bona fide *esnaf*.

Karanlık Fırın Sokak 4, off Necatıbey Caddesi (+90 212 249 6424, www.natolokantasi.com.tr)

🔔 SPLASH OUT
Çivan Er is one of the leading chefs of the so-called New Anatolian approach, in which classic Turkish plates receive a daring update. Hence

his appropriately named *Yeni Lokanta* (New Restaurant) in the south of busy Beyoğlu. Incorporating myriad meze-like starters, one main, and one dessert, tasting menus for two might have you trying raki-infused, cheese-topped sea bass. Book ahead. The godfather of New Anatolian cuisine, however, is Mehmet Gürs, who puts his Turkish-Scandinavian heritage to enterprising use at *Mikla*, which is just to the north. Request the wine-paired degustation and relish the monkfish with clams, red beans, and fig vinegar.

Kumbaracı Yokuşu 66 (+90 212 292 2550, no website); Hotel Marmara Pera, Meşrutiyet Caddesi 15 (+90 212 293 5656, www.miklarestaurant.com)

▼ Before or after any meal at Mikla, be sure to pay a visit to the venue's roof terrace, offering captivating views of the city.

QUINTESSENTIALLY ISTANBUL

Nothing's more quintessentially Turkish than sitting down to a disgustingly big platter of meze. The correct place is a meyhane (tavern), and the correct meyhane is *Asmalı Cavit*, deep in Beyoğlu. Below its low roof, tables are crammed together and askew pictures cover the walls. Start with the cold stuff: eggplant (aubergine) salads, taramasalata, marinated peppers, stuffed vine leaves, and lakerda (salted bonito) are all classics. Then go hot, perhaps with some fried mussels or thin liver slices. Wash it down with aniseed-flecked raki and just keep going until you fear explosion.

Asmalı Mescit Caddesi 16D (+90 212 292 4950, no website)

💬 LOCAL SECRET

Kardeşim Sokak is easily overlooked: at first glance, the small lane contains hardware stores, plus some more hardware stores. Persevere, though, and you'll also spy lunch-only *Tarihi Karaköy Balık Lokantası*—one of Istanbul's best grilled-fish restaurants. Today's critters are proudly displayed in a fishmonger-like front sill; behind, locals scoff shrimp skewers or sea bass cooked in baking paper. For more adventurous local eats, board the ferry to Kadıköy, on Istanbul's Asian shore, and visit *Çiya Sofrası*. The kitchen there honors forgotten peasant food, including vibrant salads and delicately spiced lamb kebabs.

▼ Çiya Sofrası offers an opportunity to eat traditonal Turkish food at an excellent price.

Kardeçim Sokak 45/A (+90 212 251 1371, no website); Güneşlibahçe Sokak 43 (+90 216 330 3190, www.ciya.com.tr)

🍽 BRILL FOR BRUNCH

"Turkish breakfast" generally means a spread: various cheeses, hard-boilcd eggs, omelet dishes with sucuk (Turkish sausage), savory pastries, olives, lots of tea, and kaymak, a clotted cream, and fresh honey. All of which are generally available at Beyoğlu's *Lades Menemen*, alongside its titular menemen: fried eggs, tomatoes, and green bell (sweet) peppers. If you'd rather have a more western brunch, visit the hipster Cihangir quarter near Taksim Square and grab a table in *Kahve6*. The traditional local breakfasts there are rivaled by mueslis, yogurts, sandwiches, and salads, all made from organic ingredients.

Sadri Alışık Sokak 11 (+90 212 249 5208, www.lades.com.tr); Anahtar Sokak 13/A (+90 212 293 0849, www.facebook.com/kahve6)

▲ Think you know köfte? Just wait until you try the Istanbul version.

🍲 REGIONAL COOKING

You won't lack for meat in Istanbul, with the scent of köfte and kebab wafting around every corner. But, as ever, the challenge comes in finding diamonds among the duds. To enjoy the finest meatballs, cross the Golden Horn to Eminönü—not far from the Spice Bazaar—and dine at family-owned *Meşhur Filibe Köftecisi*. Serving the locals for over a century, its succulent köfte are best accompanied by bread, crunchy onions, and bean salad. As for kebabs, nearby *Hamdi*

Et Lokantası combines superlative doners and urfas with terrific harbor views. Aim to reserve.

Hoca Paşa Sokak 3 (+90 212 519 3976, www.meshurfilibekoftecisi. com); Kalçın Sokak 11 (+90 212 528 0390, www.hamdi.com.tr/eminonu)

HIP & HAPPENING

Trendy Karaköy is generally the place to find buzzworthy, blogger-luring hotspots. One such is *Naif*, which provides a lighter Mediterranean alternative to the usual, rather heavy Turkish fare—right down to its bright interiors and modern art-lined walls. The seasonal menu often stars orzo pilaf with octopus, sorrel salad with avocado, lamb shanks on pasta with walnuts and Keş cheese, or even oven-baked quince with apple and clotted cream. Similarly non-native is the beautiful *SushiCo*, in whose calm, lantern-lit room hepcats wolf down crispy gingerbread chicken and nicely varied bento boxes.

Mumhane Caddesi 52
(+90 212 251 5335, www.
instagram.com/naifistanbul);
Kemankeş Caddesi 57
(+90 212 245 8801,
www.sushico.com.tr)

▲ Turkish coffee uses a finer grind than Italian versions. The grounds stay in the cup, providing different layers of texture as you drink—it starts thin and gradually gets thicker.

CAFFEINE KICKS

If it's bitter Turkish coffee you want, make for *Mandabatmaz*. Made with proper beans, the thick stuff is a ritual at this small Beyoğlu cafe and usually enjoyed with friends in slow-paced fashion, often accompanied by Turkish delight. You can sit out in the alley, under hanging plants, or in the cluttered interior. Much fluffier is the java at *Drip Coffeeist*, a micro-roaster which has branches in both Asian and European Istanbul. The latter, also in Beyoğlu, serves excellent ice-drip brew coffee using a Belgium syphon and Aeropress, and has a nice terrace.

Olivya Geçidi 1/A (+90 212 243 7737, www.mandabatmaz. com.tr); General Yazgan Sokak 9/A (+90 212 243 7552, www. facebook.com/dripcoffeeist)

SWEET TOOTH

There's baklava, and there's really good baklava. You'll find boxes of the latter at *Güllüoğlu* who, having spent 200 years making the stuff, have nailed it as one might expect. The wispy layers of phyllo dough, the punchy pistachios, the clarified butter, the heady syrup—every ingredient is perfect and perfectly measured. Buy a gift pack for someone at home and, strictly in the name of research, some for yourself. The best complement is a glass of crisp lemonade. Chocolate, walnut, gluten-free, and sushi-like rolled versions are also available.

Mumhane Caddesi 171 (+90 212 249 9680, www.karakoygulluoglu.com)

MARKET RESEARCH

After the Grand Bazaar, Istanbul's most emblematic covered market is Eminönü's beautiful, Ottoman-era *Spice Bazaar* (open daily, 8am–7:30pm)—often referred to as the *Egyptian Bazaar*. Every imaginable spice is there along the arched rows, displayed in crates and trays of eye-watering colors, and even more eye-watering smells. You'll also find lots of Turkish delight, caviar, olives, honeys, and dried fruits, while *Cankurtaran Gıda* is a store selling cheeses from across the country. Pungent tulum, made with ewe's milk in a goatskin casing, is a good shout.

Mısır Çarşısı 92 (+90 212 513 6597, www.misircarsisi.org.tr)

▼ Every conceivable spice is to be found at the Egyptian Bazaar. Pay a visit to top up your depleted spice racks back home.

Lyon

FRANCE

> Despite its (apparently undisputed) standing as "France's
> gastronomic capital" and being a mere two-hour train journey
> from Paris, the country's second city flies under the tourist radar.
> Those who do go relish pink-tinged Renaissance buildings, converging
> rivers, cathedrals, cobbles, and an unrelentingly rich diet—anchored by
> the city's endemic bouchons.

LYON

ON A BUDGET?

Near the River Sâone, *Le Potager des Halles* is a strange beast. An overhanging gallery and abundance of blonde wood lend it a saloon-type vibe. The school chairs and convivial tables scream canteen. But the loudest shouts of all come from happy eaters, most of them secluded down a spiral staircase in the dining room. Three creative lunch courses—which might include duck tartare beside creamy mushroom sauce or grilled sardines with cherries and salicornes—cost just €19. Expect to pay about €30 for dinners starring Galician steak or quince-stuffed quail.

3 Rue de la Martinière
(+33 4 72 00 2484,
www.lepotagerdeshalles.com)

CAFFEINE KICKS

Most of the regular, often ornate cafés in Lyon's center serve very bitter, palate-singeing coffee, so retreat instead to Les Cordeliers—a narrow old quarter wedged between the parallel River Rhône and River Sâone—and *Puzzle*. Minimal in design ethos, with wiry lights and mostly bare walls, this sharp coffee store's owners rotate small-roast producers from across Europe, and throw in great croissants and

▲ Paul Bocuse was a titan of French food, doing as much as anyone to certify the reputation of France as one of the world's foremost culinary nations.

lunchtime focaccias for good measure. There's a relaxing hubbub at all times, and the yellowy light is almost magnetic on rainy or chilly days.

4 Rue de la Poulaillerie
(+33 4 78 42 8872,
www.puzzlecafe.fr)

SPLASH OUT

Among Lyon's 16 Michelin star-holders is the legendary restaurant *Paul Bocuse*. Considered the godfather of nouvelle cuisine, Bocuse sadly passed away in 2018. His three-star venue in the city remains mighty tough to book. Far more feasible is snaffling a seat in one-star *Les Terrasses de Lyon* in the old town's hilly Fourvière quarter, which sits above most

of the rooftops. Lunchtime tasting menus might include duck liver with port-soaked cherries and/or calf's brain accompanied by peas and lardo, followed by a cheese trolley and chocolate mousse plus confit orange. Book to sit on the scenic terrace, if possible, and leave time to enjoy an extensive wine list.

40 Quai de la Plage, Collonges-au-Mont-d'Or (+33 4 72 42 9090, www.bocuse.fr);
25 Montée Saint-Berthélémy (+33 4 72 56 5656, www.villaflorentine.com/en/restaurant.html)

QUINTESSENTIALLY LYON

How to define a bouchon? Unstintingly old-school, these are typically small, family-run establishments with vintage pastis adverts or Marcel Pagnol film posters on the walls; their staff use chipped copper pots and serve you at rickety tables under red-checked tablecloths. Most important is a robust, modernity-mocking menu full of rich, longstanding Lyonnais dishes, plus heaps of wine— the quality is optional. Ideally, there will also be oreillers (pillows), house terrines with

▲ Look up "bouchon" in your French dictionary and you should see a picture of Chez Georges; the definitive example.

layers such as foie gras or duck heart. Nearly all of these elements are present at husband-and-wife operation *Chez Georges* in Les Cordeliers.

8 Rue du Garet (+33 4 78 28 3046, www.aupetit-bouchon-chez-georges.fr)

HIP & HAPPENING

If it's contemporary principles you want, try Lyon's new wave of young chefs and global flavors. One such is local boy Mathieu Rostaing-Tayard and his unpretentious *Café Sillon*. The carte there changes fortnightly, but the emphasis remains fixed on exotic produce and exquisite combos. Barbecued octopus, harissa, and coriander blossoms is an occasional main, as is red mullet amid a sauce of sea urchin, sesame, and star anise. The room's cozy color scheme of dark blue walls and a central farmhouse-style pine table is pleasing, as are €15 (lunch) and €38 (dinner) prix-fixe three-course offerings.

46 Avenue Jean Jaurès (+33 4 78 72 0973, www.cafe-sillon.com)

▼ Cheese displayed at Les Halles (see overleaf), plenty for even the most ardent turophile (that's a cheese lover to you and me).

 LOCAL SECRET
Hamburgers hardly seem a classically Lyonnais dish, but they actually fit in well, being hearty, heavy, and not entirely good for you. Hence the proliferation of burger bars across town. One of the local faves is *Les Frangins*, a block away from the Rhône. In fairness, the ingredients used there are locally sourced and as healthy as possible; much more importantly, each sandwich comes with unlimited frites. The double burger is bigger than most people's head, while a veggie option and yummy homemade ketchup earn further points.

9 Rue des Marronniers (+33 6 85 48 6928, www.facebook.com/ lesfranginsjr)

BRILL FOR BREAKFAST

Up in the university district near Mama Shelter's hotel is bijou *Le Kitchen Café*, and its all-day offerings. Go for lunch and you might scoff chef Connie Zagora's dried duck breast with nori or shredded pork shoulder and pumpkin. Visit later and you'll happen on Lyon's only dessert bar. Arrive anytime and pâtissier Laurent Ozan's cinnamon buns or Tonka bean and semi-sweet (dark) chocolate cookies are ready to buy. But best of all is breakfast: for those same pastries, for amazing granola, for fine fruit juices, and for the general sense of being somewhere fresh and fun.

34 Rue Chevreul
(+33 6 03 36 4275,
www.lekitchencafe.com)

REGIONAL COOKING

Lyon's location is the key point of its epicurean exceptionalism: just north of the garden-like Provence, west of the Alps, close to Burgundy and Beaujolais, on two rivers, and so on. This leads to numerous famous dishes, but none out-renown quenelles sauce Nantua or pike croquettes in crayfish broth. Have that in *Daniel & Denise*, a fancier bouchon whose quieter Créqui branch also serves gras-double à la lyonnaise—tripe, layers of sweet onion, tart—and superlative saucisson inside

▲ Quenelles sauce Nantua. It's the flavor of the crayfish sauce—this being Lyon, of course it's packed with butter—that really makes this dish stand out.

a brioche loaf. It might not be delicate, but it's darn good.

156 Rue de Créqui
(+33 4 78 60 6653,
www.danieletdenise.fr)

PERPETUAL WINE

There are more than 10,000 bottles of plonk stashed in the **Burgundy Lounge**, a two-floor temple to wine on the River Saône's market-laden east bank. You can laze upstairs amid stone walls and chunky beams, or chat with sommelier Hervé in a steel-decorated tasting room. Superb food is also available, and can be matched to wines on Saturday evenings. But the chief challenge is to try as many whites, reds, rosés, and champagnes, and buy a few too. Just make sure you don't fall overboard if riding the Vaporetto boat home.

24 Quai Saint-Antoine
(33 4 72 04 0451,
www.burgundylounge.fr)

MARKET RESEARCH

Named in tribute to, rather than because it features, the great man's creations, **Les Halles de Lyon Paul Bocuse** eschews the usual outdoor or covered format, and instead occupies a modern building beside the Part-Dieu mall. That tribute includes a higher than usual caliber of market fare: piles of crottins (goat's cheese), soft Saint Marcellin at Mère Richard, or peculiarly pear-shaped, knobbly Jésus de Lyon sausages courtesy of Charcuterie Sibilia and its charismatic female owner. The 60 stands are complemented by counter bars and restaurants serving shellfish and white-wine brunches.

102 Cours Lafayette
(+33 4 78 62 3933,
www.halles-de-lyon-paulbocuse.com)

▼ Jesus de Lyon sausages
In a city filled with excellent charcuterie, this local favorite stands out.

Paris

FRANCE

> From croissant-crammed patisseries to bistrots with garlic bulbs hanging by the door, Paris is full of culinary romance. But besides the snails and steak tartares, you'll also find terrific Asian food, cocktails, and coffee to rival anywhere, and no end of secret, insidery gems. All the more excuse to read on…

ON A BUDGET? You'll find *Bistrot Victoires* in Paris's most central 1st arrondissement, just north from the Louvre gallery. As such, its great-quality duck confit or roast chicken constitutes an absurd bargain at just over €10. But if that's €10 too much, how does free food sound? Well, kind of free. At a certain few Paris bars, ordering a drink, any drink, brings a gratis plate of nosh. One of them is *Le Grenier*, in among Oberkampf's student pubs, where the offer's

available after 8pm on Friday and Saturday evenings—usually as "gypsy jazz" is performed downstairs.

6 Rue de la Vrillière
(+33 1 42 61 4378,
no website);
152 Rue Oberkampf
(+33 1 48 05 1352,
no website)

SPLASH OUT

The City of Love hardly lacks for fine-dining, but few chefs put more amour into their work than *Pierre Gagnaire*. On record as saying he finds inspiration in jazz and paintings, Gagnaire's as liable to produce tuna meringues as Scottish grouse with citrus marmalade at his eponymous restaurant near the Arc de Triomphe. There's a geeky stress on molecular cooking here, but never at the expense of a sated stomach. Ditto in the Haut Marais at blue-colored

Les Enfants Rouges, one of Paris's many Japanese-influenced joints. Order seasoned salmon with stewed oxtail.

6 Rue Balzac (+33 1 58 36 1250; www.pierre-gagnaire. com/restaurants);
9 Rue de Beauce
(+33 1 48 87 8061,
www.les-enfants-rouges.fr)

HIP & HAPPENING

The Petite Ceinture is a disused railway which loops around the outer parts of Paris's central arrondissements. Some parts are open to walkers, and the derelict old stations are gradually reopening as cool cultural spaces. One such is Gare Ornano, in northern Paris near the Clignancourt flea market, and now become *La REcyclerie*. In the high-ceilinged space, you'll find an eco-focused brunch café, restaurant, garden, performance center, and urban farm. Dinners are locavore-

▲ La REcyclerie regularly puts on vegan and vegetarian menus, which are often tricky to find in France.

style, affordable (costing between €8–15), and always include three vegetarian options and one for vegans. Have a cocktail beside the train tracks afterward.

83 Boulevard Ornano (+33 1 42 57 5849, www.larecyclerie.com)

BRILL FOR BRUNCH

Staying in Montmartre? The best laziest brunches are devourable at *The Hardware Société*, a second café (close to the Sacré-Cœur) from Australian owners Di and Will after a debut in Melbourne. This being France, pastries are numerous—sourced from admired boulangerie Bo— and accompanied by nice jams, one of them strawberry champagne-flavored, and brioche with lime curd. Prefer savory fare? You can choose from boiled eggs beside crumbled black pudding, mini fry-ups, and crème-fraîche blinis. The decor is twee— butterfly wallpaper above light wooden furniture— and staff wear Breton stripes.

10 Rue Lamarck (+33 1 42 51 6903, www.instagram.com/hardwaresocieteparis)

▼ Café culture in Paris is sacrosanct. Hopefully this is a part of French culture that will never change.

 LOCAL SECRET
Batignolles is an affluent area just west of Montmartre, full of plane-tree squares, tempting stores, and neighborhood bistros. Best of the latter is *Gare au Gorille*, named after a song by Georges Brassens. It obediently follows the bistronomy textbook— uncomplicated but chic interiors, succinct daily menu, friendly service, informality— while throwing in an occasional hip-hop soundtrack. Unusual combos are the norm: squid with sausage and sorrel; hake and feta. The wine list is long and local, and the chocolate mousses are sensational. Five lunch courses cost €39.

68 Rue des Dames (+33 1 42 94 2402, www.gareaugorille.fr)

REGIONAL COOKING

There's traditional French fare, and then there's French fare served a third of the way up the Eiffel Tower. Now that it's overseen by the Alain Ducasse empire, *Le Jules Verne* can boast food to compare to the 400-ft (122-m) high views. All-Gallic wines from Bordeaux and Burgundy can be enjoyed with seared scallops, veal medallions in mushroom sauce, or preserved-duck foie gras

▼ Positioned in Paris's most iconic landmark, Le Jules Verne has been visited by everyone from Hollywood royalty to political world leaders.

and walnut jus. The chocolate bolt dessert pays homage to the 2.5 million bolts which hold the famous tower together. Note that you must dress smartly, and reserving is sensible.

Tour Eiffel (+33 1 45 55 6144, www.lejulesverne-paris.com)

QUINTESSENTIALLY PARIS

But if it's real French cooking you want, and the ultimate Parisian experience, then it has to be a homely bistro. There are thousands to choose from, but none are more reliable than *L'Ourcine* on the Left Bank. Set in a restored 17th-century building near the catacombs entrance, it has the perfect menu of coziness, a chatty atmosphere, wooden furniture, and simple, soulful, superb food. Order hearty game dishes such as pork mignon in the fall (autumn) and grilled cod with pea tapenade otherwise. Red wine is a must.

92 Rue Broca (+33 1 47 07 1365, www.restaurant-lourcine.fr)

🍸 KILLER COCKTAILS

Want to assist the "defense of French spirits"? That's the mission of Bonne Nouvelle's speakeasy-style *Le Syndicat*, which purports to stock only rare alcohols from "l'Hexagone" (i.e. France) and overseas territories—the likes of pomace brandy, byrrh, and numerous fruit-infused eaus de vie—and stands out amid Paris's crowded cocktail scenery. Over in the trendy 11th, *A la Française* is equally jingoistic across its two floors. Upstairs is a bistro, but head below for bartender Stephen's long-lost Gallic "coquetels" such as the "Emile Lefeuvre:" an intriguing aperitif fusing vermouth and Breton triple sec.

51 Rue du Faubourg Saint-Denis (+33 9 86 26 2472, www.syndicatcocktailclub. com); 50 Rue Léon Frot (+33 9 82 49 0269, www.alafrancaiseparis11. lafourchette.rest)

☕ CAFFEINE KICKS

Third-wave coffee bars are myriad across Paris, especially around Canal St Martin as it turns north-east. Less of a hotspot is the Bas du Dix-Huitième, or "bottom of the 18th," villagey streets safely north of Pigalle's sex shops and clip bars. But it's there you'll find alabaster-tiled, white-tiled

Café Pimpin, whose pretty lattes will fuel your walk uphill toward the Sacré-Coeur. If the coffee quality is all-important, journey to Belleville and its *Brûlerie de Belleville*. Roasted on site, the buyable beans come from small-sized producers—and Saturday mornings see hour-long coffee "dégustations," including bags to take home.

64 Rue Ramey (+33 1 46 06 9725, no website); 10 Rue Pradier (+33 9 83 75 6080, www.cafesbelleville.com)

MARKET RESEARCH

Neither the French Revolution nor the Siege of Paris could derail the city's *Marché d'Aligre* (Tues–Sat, 9am–1pm and 4pm–7.30pm; Sun, 9am–1.30pm), close to Bastille. Ignore the main yard's African masks and rare books, and head instead through the covered Beauvau section for vintage fishmonger and butcher stalls. There's fruit and veg along the street, too. Over on the Left Bank, well-to-do Place Monge hosts the *Marché Monge*'s artisan and expensive bio stalls—fine honeys, cultivated sausages—around its trickling fountain on the Latin Quarter fringes (market open Wed and Fri, 7am–2.30pm; Sun, 7am–3pm).

▲ Byrrh, a wine-based aperitif that's just one of the traditional French liqueurs found on Le Syndicat's patriotic menu.

Place d'Aligre (+33 1 85 11 0458, http://marchedaligre. free.fr); Place Monge (+33 1 48 85 9330, no website)

Bordeaux

FRANCE

> Ah, the City of Wine! But what about its food? Well, good news: in recent years, Bordeaux's grub has been revitalized, and is now on a par with its grapes. Energized by the arrival of celebrity chefs like our own Gordon Ramsay, today "Little Paris" boasts nationally admired bistros and a new wave of wine bars.

💲 ON A BUDGET?

Deep in the slumberous St. Pierre quarter, *Le Chien de Pavlov* performs a slow seduction on its patrons. First, it impresses with tiny Tiffany lights and antique furniture amid a cozy interior. Then, from the seasonally changing menu, spring forth terrific and flavor-bursting dishes which you won't find anywhere else: crab and makrut-lime ravioli, perhaps, or ginger-garnished veal chops with salsify. A predictably excellent wine list earns further favor. But the clincher is the check (bill): tipples turn out to cost from a mere €2 (with an entrée) and mains from just €13. Same time tomorrow?

45–47 Rue de la Devise (+33 5 56 48 2671, www.lechiendepavlov.com)

SPLASH OUT

There's no shortage of fine-dining in Bordeaux; hardly a surprise given all the swanky wineries nearby. One turn-up, however, is that the most in-demand venue isn't Gordon Ramsay's *Le Pressoir d'Argent* on the focal Place de la Comédie, nor indeed his French counterpart Philippe Etchebest's *Le Quatrième Mur* opposite. In fact, it's *Garopapilles*, a bijou and contemporary French restaurant west of center. The prix-fixe menu alters frequently, but you can expect such wonders as veal fillet with poached pears, cockles, and squid-ink gnocchi. Sit in the herb-scented garden, if possible.

62 Rue de l'Abbé de l'Epée
(+33 9 72 45 5536,
www.garopapilles.com)

HIP & HAPPENING

Something of a flagship for the city's dynamic modern bistro trend, low-lit *Miles* impressively bagged "Best Restaurant of 2015" from French foodie bible *Le Fooding* after just one year. Its name refers to the distance its four youthful chefs—who hail from Israel, Japan, New Caledonia, and France, but via Vietnam—covered to team up on a cobbled St. Pierre street. Such provenance is reflected on the open kitchen's menu, too, as per offerings such as veal tartare beside sesame seed oil-marinated egg yolk and swordfish with Madras curry jelly plus coconut and lime gremolata.

33 Rue du Cancera
(+33 5 56 81 1824,
www.restaurantmiles.com)

BRILL FOR BREAKFAST

Cagettes are the open wooden crates that carry fresh produce to food markets each morning; an appropriate inspiration for Théo Saint Martin's wholesome, locavore establishment. *La Cagette* is open all day, but its prix-fixe, two-course brunch/lunch—served from 11:30am until 3:30pm—is the carte to aim for. Perch on flea-market chairs and choose between scrambled eggs with asparagus, roasted potatoes, and bacon; Tomme cheese and green onion quiche; or slices of yummy meatloaf and gravy. There are homemade strawberry yogurts for dessert and fruit juices or organic rosés to sip.

8 Place du Palais
(+33 9 80 53 8435,
www.lacagette.com)

▼ Le Pressoir d'Argent offers fine dining in an opulent setting. The smartest foodies will also head to Garopapilles.

REGIONAL COOKING

Lamprey in red-wine sauce. Grilled duck breasts. Cassoulet. Pasta with cream, foie gras, ceps, and bacon. These are just a few examples of the definitively Sud Ouest cuisine on offer at *La Tupina* since 1968. Pricey without being pretentious, this institution combines its provincial meals with a provincial look: the various rooms' open hearth, hanging pots and pans, rustic furniture, and a hanging cauldron—for which tupina is the Basque word—all scream country farmhouse, despite the city-center premises. For dessert, order Bordeaux's little custard cake, cannelé, funked up here with a homemade ice-cream filling and salted-caramel drizzle.

6 Rue Porte de la Monnaie (+33 5 56 91 5637, www.latupina.com)

QUINTESSENTIALLY BORDEAUX

Cannelé is also gobbled by the customers of *Chez Pompon*, but this is just one reason to head to this century-old brasserie near broad Place des Quinconces. For the Bordelais locals, the chief one is Thursday evening's Afterwork, an often-themed affair when cut-price cocktails fuel a lively dance floor. On Fridays there are special tapas. Otherwise, it's just a dependable French brasserie in the great tradition: friendly veteran waiters, good food, cabinets packed with wine, an outside terrace, relaxed convivial atmosphere. There might as well be garlic bulbs hanging from the walls.

4 Cours de Verdun (+33 5 56 79 1313, www.chez-pompon.fr)

PERPETUAL WINE

The coolest new wine bars in Bordeaux include *Aux Quatre Coins du Vins*, where a magnetic card enables you to self-serve grand cru wines, and *Le Flacon*, whose by-the-glass potions come with a superlative food menu, including tuna rillettes with ponzu lime. But for a wonderful blend of new and old, head to the Chartrons district, whose old warehouses host antique stores and *Le Verre Ô Vin*. Its old cellar has been lent monochrome prints and hanging lights, while the bumper bottle selection is relayed by knowledgeable staff. La Cite du Vin—a giant, glass-shaped wine museum—is just to the north.

43 Rue Borie (+33 5 56 02 5209, www.bar-a-vin-bordeaux.com)

▼ Cannelés have been enjoyed in Bordeaux for centuries. Made in fluted molds to give them their distinctive shape, the pastries are flavored with rum and vanilla.

CAFFEINE KICKS

Not to be confused with a cocktail bar of the same name, *L'Alchimiste* started life solely as a roastery, utilizing a former military barracks on the Garonne's quieter right bank. In 2016, however, sensing a gap in the market, owner Arthur Audibert opened up a corresponding café in Bordeaux's cobbled downtown. Initially, it apes many cool brew bars around the world, courtesy of a stark, white-walled interior and repurposed wooden crates; there's also a tasting room with exotic tropical wallpaper. But this is France, after all, and there simply must be croissants—sourced fresh from a bakery across the street.

12 Rue de la Vieille Tour (+33 9 86 48 3793, www.alchimiste-cafes.com)

MARKET RESEARCH

The city's main food market is covered *Les Capucins*, located in Capucins Victoire. This so-called "stomach of Bordeaux" takes place every morning bar Monday. You'll see chefs there early in the morning—having passionate discussions (read: rows), as they barter with stallholders over tomatoes or tuna— and many a local. The only

▲ The daring architecture of La Cité du Vin eschews the traditions of the region's winemakers in favor of something strikingly modern. Suffice to say, it has divided opinions.

downside is that, for all its authenticity and flowers, *Les Capucins* ain't the prettiest. Much more esthetically pleasing is Chartrons' riverfront *Marché des Quais*. The thing to do there is mill a bit, stop for a plate of oysters and wine, and then mill some more.

Cours de la Marne (+33 5 56 92 2629, www.marchedescapucins. com); 142 Quai des Chartrons (no phone, www.bordeaux.fr/ o2333)

LOCAL SECRET

There are various reasons why you might miss *La Salle à Manger des Chartrons*. It's tucked away down a quiet backstreet and boasts plain, un-shouty frontage. And it's only open

for lunch from Tuesday to Thursday, plus brunch on the month's first Sunday between October and April. Run by an art restorer—hence the cool stone walls' purchasable paintings—*La Salle*'s hand-written, slate-chalked menu lists three appetizers, mains, and desserts respectively. Ingredients hail from that same Marché des Quais (see left), inspiring simple but delicious plates such as salmon gravadlax with dill chantilly cream or veal fillets in hearty morel sauce.

18 Rue Saint-Joseph (+33 6 10 01 1877, www.salle-a-manger-bordeaux.fr)

San Sebastian

SPAIN

> Home to more Michelin-starred restaurants per capita than anywhere else, Donostia—to use the local name—is also known for its legion of pintxos (Basque tapas) bars, both old- and new-school. Throw in an excellent city beach and there's no excuse for not being on the next plane out.

ON A BUDGET?

The most clever-clogs dining is undertaken at the *Basque Culinary Center*: a training school south of the center created by seven top local chefs, including Juan Mari Arzak and Martín Berasategui. As well as one-day courses, degrees, and Masters courses, there's a predictably brilliant (and stylish) cafeteria where tomorrow's cooking wizards serve €24 tasting menus, with forward-thinking wonders, such as *bacalao* (salt cod) risotto with

▼ At the Basque Culinary Center (top) you can study for a degree in gastronomy and culinary arts and serve your creations to guests at the center's own restaurant (bottom).

cilantro (coriander) salsa, in an attractively glass-walled room. You must reserve in advance, and can only do so a month ahead. Post-feed, guests are expected to fill in short reviews, and grade the students' gastronomic efforts

Paseo de Juan Avelino Barriola 101 (+34 943 57 4500, www.bculinaryclub.com)

SPLASH OUT

There are lots of places, but really there's only one place. A perennial resident of the "World's 50 Best Restaurants" list and triple-Michelin-starred, San Sebastian's greatest fine-dining haunt is *Arzak*. Juan Marí and Elena raid regional fodder to prepare scintillating menus— either à-la-carte or seven-course degustations. Look out for the signature Red Space Egg: peppers, pig trotters, mushrooms, and sea bass served, remarkably, atop a working tablet which screens sea images. Located just east of the city center, Arzak requires booking a few months

▼ Juan Marí and Elena Arzak, whose restaurant helped to bring San Sebastian to the attention of the outside world.

▲ Zeruko's "fried egg," is in fact a lemon mousse where the "yolk" is a passionfruit sphere that pops in your mouth.

ahead; watch out for late-September's film festival.

Alcalde J. Elosegi Hiribidea 273 (+34 943 27 8465, www.arzak.es)

HIP & HAPPENING

Among the cool, modern pintxo bars, *A Fuego Negro* draws most column inches—but column inches equal crowds. *Bar Zeruko* is just as good, if not better, and easier to visit. Its brand of molecular gastronomy leads to atypical dishes: pistachio-dressed balls of morcilla and foie, say, or sherry-sponsored cannelloni plus toadstool paté, pumpkin toast, and browned Idiazabal cheese. As can be the case with boundary-pushers

like this, not every dish works—but those that do are utterly memorable. Ditto the vivid green walls and vintage trestle tables.

Arrandegi Kalea 10 (+34 943 42 3451, www.barzeruko.com)

LOCAL SECRET

Narrow *Bar Sport* can resemble a sports bar mainly because, true to its name, it is a sports bar—at least in the sense of screening live sport on TVs to bellowing, fixated fans. But this is also a part-deception which leads many diners to walk elsewhere on weekends, what with soccer and fine food rarely being happy bedfellows. They are there, though: as Real Sociedad

▲ Don't let the name mislead you, Bar Sport may be showing a game on TV, but it's a sideshow to the main event: pintxos.

strive to score, you can be gobbling traditional, beautifully assembled pintxos—lamb sweetbreads or foie gras—at fabulous prices. The rough-hewn local apple cider, sagardoa, is also present and correct.

Fermin Calbeton Kalea 10 (+34 943 42 6888, www.facebook.com/barsportdonostia)

 BRILL FOR BREAKFAST

Few establishments in the Parte Vieja (stony-streeted old town) open before midday, let alone on Sundays, so already *Maiatza* is a standout for welcoming visitors daily from 9am. But its brunches aren't just available; they're superb. Beside the ace omelets—this is Spain, after all—there are fruit salads, smoothies, granola, shakshuka-style baked eggs, avocado-sprinkled bruschetta, gluten-free toast, and a range of pastries. The vibe is relaxing,

too, assuming you can snag one of the six tables or counter seats, and there's wi-fi and English-speaking staff.

Enbeltran Kalea 1 (+34 943 43 0600, no website)

🍷 PERPETUAL WINE

Over 700 wines are available at *Essencia*, from local tipples like the dry white txakoli to stuff from the Loire and Georgia, including more than 20 by the glass. Champagne and Burgundy are particular focus areas, and ditto sherry: handpicked manzanillas, amontillados, finos, and olorosos can all be sampled. The sprawling building, near to the impressive Kursaal concert hall, also contains a restaurant serving fare such as Euskal Txerri ribs or Andalusian-style fried squid at surprisingly low prices. Pairing advice is readily available. It's a decent walk from the usual downtown haunts, but well worth the effort.

Zabaleta Kalea 42 (+34 943 32 6915, www.facebook.com/ essenciawinebar)

☑ QUINTESSENTIALLY SAN SEBASTIAN

Calle 31 de Agosto is pintxos central, with an incredible concentration of bars. Those include *Bar Martinez*, resolutely vintage in atmosphere and approach. Their chard stems with ham and cheese take some beating. Across the river is the surfer Gros barrio and *Bergara*— arguably the city's most

▲ A pintxos made with slices of exceptional Spanish *jamon* and goat's cheese on slices of bread grilled on *la plancha*.

revered gastrobar, and uniquely white-walled into the bargain. If it's not on the degustation menu (six pintxos, a drink, and a dessert for €23) and if you can pronounce it, order hot txalupa (*cha-loopa*): mushrooms, shrimp, cream, and cava in puff pastry beneath grated cheese.

Calle 31 de Agosto 13 (+34 943 42 4965, www.barmartinezdonosti.com); Calle del General Artetxe 8 (+34 943 27 5026, www.pinchosbergara.es)

🍸 KILLER COCKTAILS

Cocktails are never better than when sipped by the sea. Positioned a block from La Concha's lovely crescent beach, *La Madame* is primarily a lounge restaurant, but one where the award-winning libations outstrip everything else. There's an American emphasis on show, seen in avant-garde creations like the Brown Derby: bourbon, grapefruit juice, and honey syrup. More European in mentality is a Puente, blending tequila and martini blanco with elderflower and more grapefruit. Quality ales are also offered—craft brewing is alive and well in the city. To further swot up on the city's excellent cocktail scene, spend some time consulting local food-and-drink blog Blank Palate (www.travelcookeat.com).

Calle San Bartolomé 35
(+34 943 44 4269, www.
lamadamesansebastian.com)

CAFFEINE KICKS

Five. That's how many times Javier Garcia has triumphed at the annual "Spanish Barista Championship"—making stops at *Sakona Coffee Roasters* non-negotiable for any cafecionado visiting San Sebastian. Using carefully chosen beans pre-roasted by Probat at Javier's atelier in Irún—his hometown—this bright and big café brews up classic offerings beside the River Urumea. There are counter and table seats, and bags to buy, amid nude white walls and the wooden bar. Each single-origin blend rotates; long-term, Javier plots to start his own sustainable brand.

Bajo (back), Ramón María Lili Pasealekua 2
(+34 943 04 6457,
www.sakonacoffee.com)

MARKET RESEARCH

Named after a breach in the long-vanished city wall, *La Bretxa* (daily, from 8am) sits inside a neoclassical stone-and-iron building in Parte Vieja. Inside, although it has been downsized to 40 stalls beside a mall and a Lidl, the market remains home to piquant chorizo, crumbly cheeses, and weird-looking *percebes* (gooseneck barnacles). The latter are coinsidered local delicacies and are perilously foraged in frothy, rocky coves. Adjacent Bar Azkena serves top-notch tortillas to the sore of foot. For a posher, primmer market, head south and peruse artisan delicacies at the chic *Mercado San Martin* (Mon–Sat, 8am–8pm).

Plaza La Bretxa (+34 943 43 0336, www.cclabretxa.com); Calle Urbieta 9 (no phone, www.msanmartin.es)

▼ The Basque country is one of Spain's most fertile regions. This is reflected in the abundance of items sold at La Bretxa.

Barcelona

SPAIN

> Just like the famous FC Barcelona, the city's gastronomy is pioneering and soaked in style. You can be fine-dining on tapas one night, then scoffing seafood by Barceloneta beach the next. Then there's cava to drink, markets to trawl, and a coffee scene to infiltrate. That weekend you had planned? Best make it a week.

ON A BUDGET?

Tapas always seems like an excellent budget suggestion, until the eventual bill ends up totaling €40. The trick is to find a place which serves decent portions of filling fodder. Step up *Bar Celta Pulperia*. One of many Barcelona eateries focused on food from Galicia—Spain's northwestern region—it does chunky tortillas, thick squid rings, and a house tapa of octopus dusted with paprika. Wash it all down with bowls of Ribeira wine. If the original Gothic Quarter venue (address below) is too busy, head east to a sister bar on Carrer de la Princesa in La Ribera.

Carrer de Simó Oller 3 (+34 93 315 0006, www.barcelta.com)

SPLASH OUT

After closing their award-hoovering restaurant elBulli on the coast near the French border, brothers Ferran and Albert Adrià have opened the tapas cathedral of *Tickets* in Barcelona. It is mighty tough to book, but you have a fair chance if you get yourself organized. Reservations are available via the website two

The Adria brothers' Tickets serves tapas, but you can forget ideas of patatas bravas and tortilla. This is cutting edge.

months ahead at midnight, Spanish time; 12:02am will probably be too late so be punctual. Aim for less in-demand slots, and practice in advance. Those who are lucky enough to get in will experience a sort of small-plate heaven: five separate bar-kitchens control different menu sections, and dishes range from octopus in kimchi mayonnaise to an "airbag" of Manchego cheese.

Avinguda del Paral·lel 164 (+34 93 292 4253, www.ticketsbar.es)

HIP & HAPPENING

The chefs and waiters at *Nobook* wear jail-style jumpsuits. Porque? Because they're all "prisoners of their cooking passion." If you can forgive this slightly cringey joke, there's lots to like about this street food-inspired joint, part of a glut of buzz-worthy restaurants springing up in L'Eixample. The 16-dish menu's a wacky mash-up of ingredients and styles from across the globe: Chilean lamb legs atop pumpkin and banana, tikka masalas with quail, and soft-shell crab burgers below Cajun sauce, all supported by superb cocktails. Bright orange walls and an open, metallic kitchen lend a surreal spaceship-esque feel. Despite the name, you can book ahead.

Calle de Provenza 310 bis (no phone, www.nobook.es)

LOCAL SECRET

There are three near-certainties when it comes to restaurants around Barceloneta beach: they'll be tiny, they'll be crammed with Catalans, and they'll serve seafood paella (never with bits of chorizo included, that's considered sacrilege), which hails from down the coast in Valencla and uses devastatingly fresh, caught-that-day fare. You're advised to arrive ahead of peak time—say 7pm—and wear comfy shoes, just in case. Particularly good and handily located between the seaside and Metro station is bustling **Bar Bitácora**. The paella

there is accompanied by shoulders of lamb, patatas bravas, and chocolate puddings, and three courses generally cost south of €20. Toast your good fortune on the semi-hidden rear terrace.

Carrer de Balboa 1 (+34 93 319 1110, www.facebook.com/ bitacorabarccloneta)

BRILL FOR BRUNCH

Federal's interiors are a mature hipster's dream: lots of space, retro furniture, hanging lights, plants, and straight lines.

▲ You're in Spain so you want to eat paella? Then be sure to head down to Barceloneta beach.

You can sit on a broad, ground-floor communal table, on bench seats one level up, or on the sunny roof terrace overlooking gentrified Sant Antoni. The music is slow, and the service fast. Most importantly, the brunch menu goes on forever: there are cereals and croissants, hangover-blasting juices, and heaps of hot options. Huevos al horno (baked eggs) is the local pick, but Benedicts and Florentines are available too, and are just as good. There's also a newer, equally eye-catching sister venue in the Gothic Quarter.

Carrer del Parlament 39 (+34 93 187 3607, www.federalcafe.es/barcelona)

REGIONAL COOKING

You'll find Catalan classics such as escalivada (chargrilled eggplant/aubergine, roasted red bell/sweet peppers, and onions served cold with anchovies) and esqueixada (salt-cod salad) amid an all-Iberian menu at *La Vinateria del Call*. You'll find lots of snack plates, too, from artisanal cheeses to cider-cooked chorizo. You'll find top-notch desserts, none better than the homemade fig ice cream. You'll find an extensive wine list. You'll find a lovely locale: El Call, the higgledy, small-laned, Jewish portion of the Gothic Quarter. And you'll find, on arrival, a large and offputting telly at the door. But ignore this—you'll find that to be a very good decision.

Carrer de Sant Domènec del Call 9 (+34 93 302 6092, www.lavinateriadelcall.com)

▼ The twisting, turning streets of El Call can be disorientating. Leave plenty of time to make it to your dinner reservation.

QUINTESSENTIALLY BARCELONA

First, a confession: *La Vinya del Senyor*, as the name suggests, is more wine bar than restaurant. It does serve quality tapas, though, and boy will you need them. Marvelously located on a plaza beneath beautiful Santa Maria del Mar—the people's church—this poky den has a list of tipples longer than Don Cervantes. In particular, it's a smart place to try Catalonia's specialty, cava, which, like champagne, is made without aggressive carbonization. By-the-glass options from boutique wineries are available. The small plates run to smoked salmon, walnut rolls, and cheese plates; eat more, and you can drink more.

Plaça Santa Maria 5 (+34 93 310 3379, www.facebook.com/ vinyadelsenyor)

▲ Situated on the 11th floor, the setting of Barceló Ravel Hotel's 360° bar is only matched by its quality cocktails.

KILLER COCKTAILS

You usually need a password to access Dry Martini's *SPEAKEASY*—through the kitchen from its main bar—as part of a 1920s vibe. Think white-jacketed barmen, wood-paneling, and vintage bottles. The beautifully made drinks encompass truffle-infused vodkas and the local vermouth, but, ultimately, there's only one respectable order: a Dry Martini, of course. Keeping track of how many have been made thus far, the counter has already passed one million. If you'd prefer somewhere brighter, head for the *Barceló Raval Hotel* and the city's premier rooftop terrace bar: *360°* has panoramic views, to back up its name, and an extensive cocktail list. There's a solarium to top up your tan and a pool to cool off in. If that's not enough, live DJs play every Thursday, Friday, and Saturday.

Carrer Aribau 162 (+34 93 217 5072, www.drymartiniorg.com); Rambla del Raval 17–21 (+34 93 320 1490, www.barcelo.com)

CAFFEINE KICKS

Great brew bars aren't hard to find in Barcelona: *Nomad*, *Slice of Life*, and hipster *Satan's Coffee Corner* all impress. But *SlowMov,* secreted in the attractive Gràcia Barrio beside Gaudí's remarkable Park Güell, boasts the most impressive principles. Because owner Carmen Callizo doesn't just make superlative filtered, espresso, and cold-brew java using Arabica from Paris's Coutume Café—where she trained under Antoine Netien—in an attractive, flower-filled old workshop. Nope: she also energetically promotes slow food, and sells local producers' jellies (jams), oils, vegetables, and beers.

Carrer de Luis Antúnez 18 (+34 93 667 2715, www.slowmov.com)

MARKET RESEARCH

There are food markets and then there's *La Boqueria*. Attracting some 45,000 visitors per day, this indoor behemoth on La Rambla is a kaleidoscopic helter-skelter of smells, sounds, and sights, which sells everything your stomach can think of: seafood, sausages, leeks, lemons, cheeses, chives, chocolate, and more, much more. Allegedly in existence since 1217, the *Mercat*'s best asset is how, despite being a tourist attraction, it has also remained a local institution. You'll see housewives shopping and chefs perusing. There are places to eat, too, and chance to try peus de porc (pig's trotters) stewed with snails, or garlicky percebes (goose-necked barnacles). The market is closed on Sundays.

La Rambla 91 (+34 93 318 2584, www.boqueria.info)

▼ Be sure to vist La Boqueria early to avoid the crowds.

Madrid

SPAIN

> Food is a crucial part of Madrileño life. It's a cliché, but then aren't the best clichés always true? Along with world-class galleries, superb shopping, hipster areas, gardens, and the famous nightlife, one constant in the Spanish capital is the sight of people out eating or drinking—permanently overflowing from the many restaurants, bars, and bodegas.

ON A BUDGET?

Found inside the gastronomically revived Literary Quarter (aka Huertas) and its genteel, tree-lined streets, *La Sanabresa* is a traditionally rooted restaurant known for exceptionally low tariffs. The *menu del día* costs €16 for three courses plus wine (or €14 with water), including mains such as slow-cooked lamb leg in carrot broth or fried eggplant (aubergine). Service can be grumpy and you shouldn't expect Michelin-level fodder—but, for the price,

▲ Elaborate paintings on tiled walls are a feature found in restaurants across Madrid.

▲ The Casino de Madrid has been around since 1836 and moved into its current home on Calle de Alcalá in 1910.

it's a standout. Advance bookings for four or more are usually accepted—they speak a little English, with translations on menus.

Calle Amor de Dios 12
(+34 91 429 0338, no website)

SPLASH OUT

If you can resist the lure of a three-Michelin-star restaurant (*DiverXO*) run by a mohawk-sporting chef slash enfant terrible (David Muñoz), turn instead to two-star *La Terraza del Casino*. The reasons to do this? Wacky tapas, such as frozen gazpacho rocks with royal crab and "false squid" risotto with pink

grapefruit and eucalyptus. The presence of one Ferran Adrià as a consultant. A glorious location atop the Casino de Madrid, a neoclassical private members' club rather than a gambling den. And that seductive, titular rooftop terrace, regally overlooking a swathe of Madrid. Book well ahead.

Calle de Alcalá 15
(+34 91 532 1275,
www.casinodemadrid.es)

HIP & HAPPENING

Hiding on the fringes of ritzy Salamanca, a stone's throw from the Plaza de Colón, *Pink Monkey* promises to fuse Asian cooking with Peruvian, Mexican, and Mediterranean influences. Above all, it promises fun: oxtail dumplings and foie gras, rice pasta with langoustines and chipotle, crab bao, and, most popular of all, Thai-influenced ceviche with Alaskan salmon. The interiors are just as fun— a cheery clutter of pot plants and checker-tiled pillars, inside which sit a young and handsome crowd. Gingery cocktails help too, as do brioche-cake puddings.

Calle del Monte Esquinza 15
(+34 91 310 5272, www.
restaurantepinkmonkey.com)

BRILL FOR BRUNCH

Continue west from Pink Monkey and you'll enter hipster district Chamberí. Those bearded, designer-looking types need to eat, of course, and many do so of a late morning at *The Toast Café*. Madrid's best brunch menu spans toasties, croissants, eggs Benedict, omelets, salmon and cream cheese bagels… nothing especially original, then, but the quality is high and *Toast*'s warm decor and soothing ambience is wholly seductive. Same goes for the circa-€15 checks (bills). Bloody Marys are available for anyone in a hair-of-the-dog mood, along with Bucks Fizz.

Calle Fernando el Católico 50 (+34 91 549 3802, www.thetoastcafe.es)

💬 LOCAL SECRET

Nakeima is definitely a local secret—how many tourists visit Madrid aiming to check out a self-proclaimed "dumpling bar"?—but a lot of locals sure do know about it. So many, in fact, that you generally need to show up two hours in advance, get a ticket, and wait it out in a nearby bar. Lucky that Chamberí boasts plenty of those, too. But why so popular? Because Nakeima does much more than dumplings: Iberian-ham nigiri, pork-jowl siu mai, chili crabs, fish-butter tataki, and red-cooked pork belly are but a few of its exquisite offerings.

Calle de Meléndez Valdés 54 (+34 62 070 9399, www.twitter.com/nakeima)

🍲 REGIONAL COOKING

According to *Guinness World Records*, who tend to know about these things, **Sobrino de Botín** near Plaza Mayor is the world's oldest restaurant. It definitely dates back to 1725,

▼ Roast pork is the star dish at Sobrino de Botín, with around 50 suckling pigs roasted there each day.

and definitely featured in Hemingway's *The Sun Also Rises*. Today its gorgeous interiors and extensive wine tunnels still impress, as does the roast suckling pig. Equally venerable is Madrid's famous *bocadillo de calamares* sandwich. Nowhere fries more perfectly golden squid than rustic bar **Celso y Manolo**, to the north, where each crunchy baguette also gets slathered in lemon aïoli. Yum, yum, and yum again.

Calle de los Cuchilleros 17 (+34 91 366 4217, www.botin.es); Calle de la Libertad 1 (+34 91 531 8079, www.celsoymanolo.es)

QUINTESSENTIALLY MADRID

Forget tapas bars; in Madrid it's all about *casas de comidas*— traditional, humble taverns serving old timer Castilian dishes for not very much at all. One of the most authentic is **Casa Salvador** in vibrant, youthful Chueca. The walls contain paintings of bullfighters, photos of bullfighters, or photos of people who liked bullfighters, from Sophia Loren to Ernest Hemingway (him again). As for the menu, that runs to veal meatballs, bean-and-chorizo soup, and hake croquettes, all accompanied with warm

▲ Jamon ibérico de bellota is ham made from free-range pigs fed on acorns. It's considered the Champagne of hams.

homemade bread and served by waiters in '60s garb. Try to save some room for the nougat ice cream.

Calle Barbieri 12 (+34 91 521 4524, www.casasalvadormadrid.com)

KILLER COCKTAILS

Staying in Chueca, head to Plaza de Chueca and join the locals with a vermouth at aperitivo time (early evening). Along with wood paneling and flamboyant lamp-lights, the *Taberna de Ángel Sierra* has the fortified wine on tap. Once things get rowdy and spill outside, journey south a couple of centuries (and half a mile) to *Salmon Guru*. Celebrity bartender Diego Cabrera's latest *coctelería* pours classics and new concoctions amid retro neon signs: try a Chipotle Chillón if you can cope with mezcal and chili in tandem.

▲ Taberna de Ángel Sierra celebrated its 100th birthday in 2017 and remains resolutely traditional in its appearance.

Calle de Gravina 11 (+34 91 531 0126, www.facebook.com/tabernadeangelsierra); Calle de Echegaray 21 (+34 91 000 6185, www.facebook.com/salmonguru)

CAFFEINE KICKS

Two options for you here: one in the original trendy barrio and another in its successor. The former is Malasaña's *Toma Café*,

Madrid's first proper craft coffee shop, where offerings range from classic espresso pulls to overnight cold brews, with Aeropress plungers at the ready. The most adventurous order is an espresso-tonic—which is as it sounds—while free wi-fi and warm interiors make lingering a necessity. The newer kid on the newer block is *Hola Coffee*, over in multicultural, yoga-loving Lavapiés—exposed stone and pale wood, and drinks by national barista champion Pablo Caballero.

Calle de la Palma 49
(+34 91 702 5620,
www.tomacafe.es);
Calle del Dr. Fourquet 33
(+34 91 056 8263,
www.hola.coffee)

 MARKET RESEARCH

If you really, really like seafood, there are fewer better places on land than the Mercado de Pescados. Sure, the nearest coastline is some 200 miles away, but this is still the world's fourth biggest fish market, with over 140,000 tonnes sold each year. It's a bit one-dimensional for visitors, however—although adjacent seafood bar La Lonja del Mar is excellent—so steer instead for the *Mercado de San Miguel* (open daily: Mon–Wed and Sun, 10am–midnight; Thurs–Sat, 10am–2am). Reopened in 2009, this central affair constantly impresses, from its wrought-iron frontage to saliva-inducing displays of produce to the Beer House, who serve *cañas* (small beers).

Plaza de San Miguel
(+34 91 542 4936,
www.mercadodesanmiguel.es)

▼ All key food groups are covered at Mercado de San Miguel, including booze, if that can be considered a food group.

Porto

PORTUGAL

> Flanked by Douro vineyards, the Atlantic, and game-happy Trás-os-Montes, Portugal's second city was never going to lack for food. Join Porto's locals in traditional tascas and tavernas —some of them offering incredible value—along blue-tiled lanes, or sip pink port-and-tonics with trendies at elevated riverside perches.

ON A BUDGET?
Bombarda (as it's generally known) is a revived easterly art district, rich in contemporary galleries which usually open on Saturday afternoons. It also boasts a small, cool mall in CC Bombarda (www.ccbombarda. blogspot.co.uk). There, alongside independent fashion, toy, and jewelry stores, you'll find café/restaurant *Pimenta Rosa*, where the lunchtime buffet sees a fish or meat plate, plus vegetables/potatoes/rice,

side salad, and cup of fruit cost just €4.95. After 4pm, its esteemed fruit cakes are €1.50 per portion; after 7pm, slices of pizza go for a euro. Bargain!

Rua de Miguel Bombarda 285 (+351 93 366 2289, www.facebook.com/ restaurantepimentarosa.porto)

🔔 **SPLASH OUT**
Don't let the deliberately rustic interiors, which see monochrome family-style photos tacked to cave-like stone walls under wooden beams, fool you: *O Paparico* is a luxurious temple of Portuguese cuisine. Hearty, ten-plate "Portugality" tasting menus (€120) might potentially include beef and wild mushrooms, pork tenderloin steeped in apple

sauce, lobster rissoles with fish roe, or Sctúbal-style, wine-stewed red mullet. It's all served by romantic candlelight, and supported by an immense, 1,200-bottle wine cellar covering all of Iberia. The middle-of-nowhere location is very northerly, but only 20 minutes by taxi.

Rua de Costa Cabral 2343 (+351 22 540 0548, www.opaparico.com)

🍷 **PERPETUAL WINE**
Porto is best known for port: four-fifths of its name, hailing from vineyards east up the River Douro and traditionally bottled at grand, British-run, 17th-century waterside lodges in Vila Nova de Gaia, across Ponte Dom Luís I from everything else. Tasting

▲ Tasting halls at Taylor's in Villa Nova de Gaia, the home of the port industry.

tours are offered at many of these: the ones at still-working Taylor's are generally the best. Alternatively, the *Loja do Porto* (closed at weekends), overseen by Portugal's port institute, provides a sampling option back on the north bank. Located in Cedofeita, it also offers a souvenir-dream store and sciencey laboratory tours.

Rua Ferreira Borges 27 (+351 22 207 1669, www.ivdp.pt)

LOCAL SECRET

Some tabernas are mostly populated by Tripeiros (locals) for one good reason: they lack English menus. But foreigners are still perfectly welcome at Ribeira's *Taberna Dos Mercadores*, with some friendly waiters speaking a smattering of Inglês. Seafood's the specialty there, with menus dictated by each day's catch. To flee the tourist crowds, venture to the northern beach town of Matosinhos, but veer toward its unassuming docks. There cheap *tascas* such as *Salta-Ó-Muro* chargrill and pan-fry just-snaffled sole, sea bass, and turbot with new potatoes and herbs. Simple, maybe; superb, definitely.

Rua dos Mercadores 36 (+351 22 201 0510, no website); Rua Heróis de França 386 (+351 22 938 0870, www.saltaomuro.pt)

BRILL FOR BRUNCH

Back at Bombarda, *Rosa Et Al* is both a seven-suite hotel and superb food stop. The weekend-only brunch menu in its minimal restaurant spans egg dishes from Arlington to Benedict, croques, pancakes, fruit salads, cereals, and breads galore. There's even a three-course degustation. Everything is high-quality and small-

▲ The francesinha in all its gut-busting glory. This isn't a sandwich to grab on the go, it demands attention (and cutlery).

producer; the only put-off is an over-formality which can be less welcome in the morning. More relaxed is *Zenith Brunch & Cocktails*, in postcardy Baixa, where bespoke brunch selections and Bloody Marys are enjoyed on chunky wooden tables and faux red-brick walls.

Rua do Rosário 223 (+351 91 600 0081, www.rosaetal.com); Praça de Carlos Alberto 86 (+351 22 017 1557, www.facebook.com/zenithporto)

REGIONAL COOKING

Look away now, light eaters and waif types. Porto's most famous snack is the francesinha, a gut-shuddering sandwich of steak, ham, and *linguiça* (garlicky pork sausage), covered in melted cheese—and often a fried egg—and then drenched in hot tomato and beer sauce. Try Cedofeita's no-frills *Café Santiago*. Another staple is fried octopus fillets, fresh from the sea. You'll find them at excellent *Adega São Nicolau* in Ribeira, alongside chicken bordelaise and tongue stews. The terraced terrace cutely flows uphill with a staircase, but you'll need to book well ahead to sit there.

Rua de Passos Manuel 226 (+351 22 205 5797, www.caferestaurantesantiago.com.pt); Rua de São Nicolau 1 (+351 22 200 8232, www.facebook.com/adegasnicolau)

QUINTESSENTIALLY PORTO

Small "tavernas" are ten-a-penny in Porto, making it easy for standouts to fly under the radar. Step forward family-run *Taberna de Stª António*, close to the Virtudes *mirador* (viewpoint). Go there to experience warm service from Dona Herminia and her team,

occasional birthday singalongs, and an innate coziness born of jam-packed tables and kooky heirlooms. The menu's highlights include *Bacalhau a Braga*—fried cod in paprika sauce beside fries—and chicken with blood rice. You'll be wanting to order the homemade chocolate mousse, too, before a digestive glass of *aguardente* (Portuguese grappa).

Rua das Virtudes 32 (+351 22 205 5306, www.facebook.com/tabernastoantoniovirtudes)

▼ Porto has embraced the pedestrianized street, which allows al fresco dining to thrive.

🍸 KILLER COCKTAILS

Not content with being Porto's most desirable hotel address, **The Yeatman** also hoards a two-Michelin-star restaurant and the city's prime perch on which to drink pink port-and-tonics. From its Orangerie restaurant at night, Porto twinkles far below, ribboned around the snaking, serene Douro. If that's a bit posh/pricey for your boozing taste, steer instead for **Maus Hábitos**: concert hall, restaurant, "cultural interaction" center, and avant-garde gallery in one, and thrillingly draped across a Soviet-style apartment block's fourth floor. Classic cocktails are offered alongside wine and bottles of Super Bock.

Rua do Choupelo (+351 22 013 3185, www.the-yeatman-hotel.com); Rua de Passos Manuel 178 (+351 93 720 2918, www.maushabitos.com)

☕ CAFFEINE KICKS

The old-school way to do coffee in Porto is to stop at one of its many pastelerias, and have your *cimbalino* (espresso) with a custard cake or Bola de Berlim donut. **Majestic Café**, in Cedofeita, throws in the gilt mirrors and bas-relief cherubs of an Art Deco addict's marbly dreams. The new-school approach, meanwhile, is to go to somewhere like Baixa's **Brando**, and enjoy something

▼ Fans of pastel de nata, the famous Portuguese custard tart, are well catered for at Majestic Café.

familiarly millennial: exposed stone, white walls, funky hanging lights, and specialty, small-batch beans. Everything from iced to espresso is served, as is brunch daily.

Rua de Santa Catarina 112 (+351 22 200 3887, www.cafemajestic.com); Avenida de Rodrigues de Freitas 147 (+351 91 103 8393, www.facebook.com/ brandocasadocafe)

▲ In the age of gentrification the Mercado do Bolhão pulls off the impressive trick of offering something for all comers.

 MARKET RESEARCH

Picturesquely covered in wrought iron, 200-year-old *Mercado do Bolhão* (Mon–Fri, 7am–5pm; Sat, 7am–1pm) is liveliest on Friday and Saturday mornings. In the scheme of Europe's gentrified food markets, it successfully occupies a middle ground: while rather gourmet food stalls offer sardines and Portuguese vinho to enjoy, there are still authentic vendors selling authentic food—fruit, vegetables, cheese, olives, breads, fish, and every possible cut of meat—to authentic customers. Even more authenticity is on offer at nearby Rua Sá da Bandeira. Delicatessens there teem with beans, *bacalhau* (salt cod), peppers, breads, and brines.

Rua Formosa 214 (+351 22 332 6024, no website)

Index

ACKNOWLEDGMENTS

The publishers would like to thanks the restaurants, press agencies, and tourist boards who took the time to answer questions and provide images for inclusion in the book.

PICTURE CREDITS

Key: (L) Left; (M) Middle; (R) Right; (T) Top; (B) Bottom

P1 Oleksandr Prykhodko/Alamy; **p2** PJPhoto69/Getty; **p3** Steve Baxter/Loupe Images; **p5** (L) Jan Kees Steenman/RIJKS Restaurant, (M) S Lubenow/LOOK-foto/Getty, (R) Gavin Kingcome/Loupe Images; **p8** Leon Werdinger/Alamy; **p7** (T) Sean Pavone/Alamy, (M) travelstock44/Alamy, (B) Nikada/Getty; **p8** Dhwee/Getty; **p10** Yoann JEZEQUEL Photography/Getty; **p11** Amass Restaurant/VisitDenmark; **p12** Raffaele Nicolussi/Getty Images; **p14** Julia Davila-Lampe/Getty; **p15** Thomas Steen Sørensen/VisitDenmark; **p16** Mikael Damkier/Alamy; **p17** Penny & Bill Stockholm; **p18** robertharding/Alamy; **p19** staffan_eliassonjoruku@gmail.com; **p20** John Kellerman/Alamy; **p21** Pictures Colour Library/Alamy; **p22** Aizle Restaurant; **p23** Duncan Hale-Sutton/Alamy; **p24** (L) Tony Smith/Alamy, (R) Jorg Beuge/Alamy; **p25** Iain Masterton/Alamy; **p26** John Kellerman/Alamy; **p27** Tony French/Alamy; **p28** Peter Cripps/Alamy; **p29** Jeff Gilbert/Alamy; p30 Dave Stevenson/Alamy; **p31** Chris Lawrence/Alamy; **p32** Piet De Kersgleter/visitflanders.com; **p33** Loupe Images; **p34** Milo-Proti/visitflanders.com; **p35** FALKENSTEINFOTO/Alamy; **p36** Noppasin Wongchum/Alamy; **p37** Instock Restaurant; **p38** mauritius images GmbH/Alamy; **p39** Jan-Kees Steenman/RIJKS Restaurant; **p40** Teo Krijgsman/Madam; **p41** Peter Horree/Alamy; **p42** Sean Pavone/Alamy; **p43** (T) Colin Utz Photography/Alamy, (B) imageBROKER/Alamy; **p44** McCanner/Alamy; **p45** (T) Ekaterina Smirnova/Getty, (B) katielaurend/Stockimo/Alamy; **p46** Adam Eastland/Alamy; **p47** Agencja Fotograficzna Caro/Alamy; **p48** Jerzy Stachera/Getty; **p49** Eddie Gerald/Alamy; **p50** Stary Dom Restaurant; **p51** Hala Koszyki; **p52** Miguel Sanz/Getty; **p53** Eska Restaurant; **p54** Kevin George/Alamy; **p55** Ivoha/Alamy; **p56** Rebecca Ang/Getty; **p57** travelstock44/Alamy; **p58** Simon Reddy/Alamy; **p59** Zürich Tourism/Caroline Minjolle; **p60** xavierarnau/Getty; **p61** Hackenberg-Photo-Cologne/Alamy; **p62** Westend61/Getty; **p63** imageBROKER/Alamy; **p64** Luise Berg-Ehlers/Alamy; **p65** Food and drinks/Alamy; **p66** Slovenian Tourist Board; **p67** Bjanka Kadic/Alamy; **p68** John Anthony Rizzo/Getty; **p69** Uros Poteko/Alamy; **p70** Peter Ptschelinzew/Getty; **p71** Gordana Sermek/Alamy; **p72** Alen Gurovic/Alamy; **p73** Insights/Getty; **p74** Mauritius images GmbH/Alamy; **p75** David Silverman/Getty; **p76** Danny Nebraska/Alamy; **p77** David Silverman/Getty; **p78** Massimo Pizzotti/Getty; **p79** Loupe Images; **p80** Peter Forsberg/Alamy; **p81–3** Loupe Images; **p84** Michal Krakowiak/iStock; **p85** REDA &&CO srl/Alamy; **p86** Yadid Levy/Alamy; **p87** Francesco Lorenzetti/Alamy; **p88** S Lubenow/LOOK-foto/Getty; **p89** Guy Brown/Alamy; **p90** Oleksandr Prykhodko/Alamy; **p91** Alex Archontakis/Alamy; **p92** hayatikayhan/iStock; **p93** Terry Harris/Alamy; **p94** Alexander Spatari/Getty; **p95** Rawdon Wyatt/Alamy; **p96** Hackenberg-Photo-Cologne/Alamy; **p97** Loupe Images; **p98** Michael Sugrue/Getty; **p99** Ignacio Perez Bayona/Alamy; **p100** Ixefra/Getty; **p101** Chris Hellier/Alamy; **p102** Hemis/Alamy; **p103** travelstock44/Alamy; **p104** George Oze/Alamy; **p105** Sylvie Jarrossay/Alamy; **p106** Ian Cumming/Getty; **p107** Directphoto Collection/Alamy; **p108** allOver images/Alamy; **p109** Bruno De Hogues/Getty; **p110** Directphoto Collection/Alamy; **p111** Photo 12/Alamy; **p112** Ken Welsh/Getty; **p113** Hemis/Alamy; **p114** Greg Elms/Getty; **p115** Patrick Aventurier/Getty; **p116** San Sebastián Tourism & Convention Bureau; **p117** age fotostock/Alamy; **p118** (T) age fotostock/Alamy, (B) Benedicte Desrus/Alamy; **p119** age fotostock/Alamy; **p120** James Sturcke/Alamy; **p121** jose manuel bielsa/San Sebastián Tourism & Convention Bureau; **p122** Pere Sanz/Alamy; **p123** David Ramos/Getty; **p124** Loupe Images; **p125** Peter Horree/Alamy; **p126** Robert Marquardt/Getty; **p127** Chris Mellor/Getty; **p128** Sylvain Sonnet/Getty; **p128** (L) Krzysztof Dydynski/Getty, (R) Hemis/Alamy; **p130** Alex Segre/Alamy; **p131** Luis Martinez/Design Pics/Getty; **p132** SBMR/Alamy; **p133** Bruno Ehrs/Getty; **p134** Eve Livesey/Getty; **p135** Hercules Milas/Alamy; **p136** RossHelen editorial/Alamy; **p137** Peter Charlesworth/Getty; **p138** Stefano Politi Markovina/Alamy; **p139** Simon Reddy/Alamy

 SPLASH OUT
Based on a design by Nakul Dhaka from the Noun Project

 BRILL FOR BRUNCH
Created by Mello from the Noun Project

 BRILL FOR BREAKFAST
Created by Gregor Cresnar from the Noun Project

 QUINTESSENTIALLY...
Based on a design by Gregor Cresnar from the Noun Project

 CAFFEINE KICKS
Based on a design by Viktor Minuvi from the Noun Project

MARKET RESEARCH
Based on a design by Ayesha Rana from the Noun Project

 ON A BUDGET?
Created by Icon Solid from the Noun Project

 HIP & HAPPENING
Based on a design by Artem Kovyazin from the Noun Project

 LOCAL SECRET
Created by hunotika from the Noun Project

 WORTH A TRIP
Based on a design by Wahyuntitle from the Noun Project

 REGIONAL COOKING
Based on a design by Arthur Shlain from the Noun Project

 PERPETUAL WINE
Based on a design by Alex Furguiele from the Noun Project

 KILLER COCKTAILS
Created by Jony from the Noun Project

 DEER OH BEER
Created by Paul Tilby

 FOREIGN FODDER
Based on a design by mikicon from the Noun Project

 ICE CREAM DREAM
Based on a design by Vectors Market from the Noun Project

 DELECTABLE DAY TRIP
Created by Iconfactory Team from the Noun Project

 SWEET TOOTH
Based on a design by Edward Boatman from the Noun Project

 PUB GRUB
Created by Vectors Market from the Noun Project

ACKNOWLEDGMENTS

The publishers would like to thanks the restaurants, press agencies, and tourist boards who took the time to answer questions and provide images for inclusion in the book.

PICTURE CREDITS

Key: (L) Left; (M) Middle; (R) Right; (T) Top; (B) Bottom

P1 Oleksandr Prykhodko/Alamy; p2 PJPhoto69/Getty; p3 Steve Baxter/Loupe Images; p5 (L) Jan Kees Steenman/RIJKS Restaurant, (M) S Lubenow/LOOK-foto/Getty, (R) Gavin Kingcome/Loupe Images, p8 Leon Werdinger/Alamy, p7 (T) Sean Pavone/Alamy, (M) travelstock44/Alamy, (B) Nikada/Getty; p8 Dhwee/Getty; p10 Yoann JEZEQUEL Photography/Getty; p11 Amass Restaurant/VisitDenmark; p12 Raffaele Nicolussi/Getty Images; p14 Julia Davila-Lampe/Getty; p15 Thomas Steen Sørensen/VisitDenmark; p16 Mikael Damkier/Alamy; p17 Penny & Bill Stockholm; p18 robertharding/Alamy; p19 staffan_eliassonjoruku@gmail.com; p20 John Kellerman/Alamy; p21 Pictures Colour Library/Alamy; p22 Aizle Restaurant; p23 Duncan Hale-Sutton/Alamy; p24 (L) Tony Smith/Alamy, (R) Jorg Beuge/Alamy; p25 Iain Masterton/Alamy; p26 John Kellerman/Alamy; p27 Tony French/Alamy; p28 Peter Cripps/Alamy; p29 Jeff Gilbert/Alamy; p30 Dave Stevenson/Alamy; p31 Chris Lawrence/Alamy; p32 Piet De Kersgleter/visitflanders.com; p33 Loupe Images; p34 Milo-Profi/visitflanders.com; p35 FALKENSTEINFOTO/Alamy; p36 Noppasin Wongchum/Alamy; p37 Instock Restaurant; p38 mauritius images GmbH/Alamy; p39 Jan-Kees Steenman/RIJKS Restaurant; p40 Teo Krijgsman/Madam; p41 Peter Horree/Alamy; p42 Sean Pavone/Alamy; p43 (T) Colin Utz Photography/Alamy, (B) imageBROKER/Alamy; p44 McCanner/Alamy; p45 (T) Ekaterina Smirnova/Getty, (B) katielaurend/Stockimo/Alamy; p46 Adam Eastland/Alamy; p47 Agencja Fotograficzna Caro/Alamy; p48 Jerzy Stachera/Getty; p49 Eddie Gerald/Alamy; p50 Stary Dom Restaurant; p51 Hala Koszyki; p52 Miguel Sanz/Getty; p53 Eska Restaurant; p54 Kevin George/Alamy; p55 Ivoha/Alamy; p56 Rebecca Ang/Getty; p57 travelstock44/Alamy; p58 Simon Reddy/Alamy; p59 Zürich Tourism/Caroline Minjolle; p60 xavierarnau/Getty; p61 Hackenberg-Photo-Cologne/Alamy; p62 Westend61/Getty; p63 imageBROKER/Alamy; p64 Luise Berg-Ehlers/Alamy; p65 Food and drinks/Alamy; p66 Slovenian Tourist Board; p67 Bjanka Kadic/Alamy; p68 John Anthony Rizzo/Getty; p69 Uros Poteko/Alamy; p70 Peter Ptschelinzew/Getty; p71 Gordana Sermek/Alamy; p72 Alen Gurovic/Alamy; p73 Insights/Getty; p74 Mauritius Images GmbH/Alamy; p75 David Silverman/Getty; p76 David Silverman/Getty; p77 David Silverman/Getty; p78 Massimo Pizzotti/Getty; p79 Loupe Images; p80 Peter Forsberg/Alamy; p81–3 Loupe Images; p84 Michal Krakowiak/iStock; p85 REDA &&CO srl/Alamy; p86 Yadid Levy/Alamy; p87 Francesco Lorenzetti/Alamy; p88 S Lubenow/LOOK-foto/Getty; p89 Guy Brown/Alamy; p90 Oleksandr Prykhodko/Alamy; p91 Alex Archontakis/Alamy; p92 hayatikayhan/iStock; p93 Terry Harris/Alamy; p94 Alexander Spatari/Getty; p95 Rawdon Wyatt/Alamy; p96 Hackenberg-Photo-Cologne/Alamy; p97 Loupe Images; p98 Michael Sugrue/Getty; p99 Ignacio Perez Bayona/Alamy; p100 Ixefra/Getty; p101 Chris Hellier/Alamy; p102 Hemis/Alamy; p103 travelstock44/Alamy; p104 George Oze/Alamy; p105 Sylvie Jarrossay/Alamy; p106 Ian Cumming/Getty; p107 Directphoto Collection/Alamy; p108 allOver images/Alamy; p109 Bruno De Hogues/Getty; p110 Directphoto Collection/Alamy; p111 Photo 12/Alamy; p112 Ken Welsh/Getty; p113 Hemis/Alamy; p114 Greg Elms/Getty; p115 Patrick Aventurier/Getty; p116 San Sebastián Tourism & Convention Bureau; p117 age fotostock/Alamy; p118 (T) age fotostock/Alamy, (B) Benedicte Desrus/Alamy; p119 age fotostock/Alamy; p120 James Sturcke/Alamy; p121 jose manuel blesa/San Sebastián Tourism & Convention Bureau; p122 Pere Sanz/Alamy; p123 David Ramos/Getty; p124 Loupe Images; p125 Peter Horree/Alamy; p126 Robert Marquardt/Getty; p127 Chris Mellor/Getty; p128 Sylvain Sonnet/Getty; p128 (L) Krzysztof Dydynski/Getty, (R) Hemis/Alamy; p130 Alex Segre/Alamy; p131 Luis Martinez/Design Pics/Getty, p132 SBMR/Alamy; p133 Bruno Ehrs/Getty; p134 Eve Livesey/Getty; p135 Hercules Milas/Alamy; p136 RossHelen editorial/Alamy; p137 Peter Charlesworth/Getty; p138 Stefano Politi Markovina/Alamy; p139 Simon Reddy/Alamy

SPLASH OUT
Based on a design by Nakul Dhaka from the Noun Project

BRILL FOR BRUNCH
Created by Mello from the Noun Project

BRILL FOR BREAKFAST
Created by Gregor Cresnar from the Noun Project

QUINTESSENTIALLY...
Based on a design by Gregor Cresnar from the Noun Project

CAFFEINE KICKS
Based on a design by Viktor Minuvi from the Noun Project

MARKET RESEARCH
Based on a design by Ayesha Rana from the Noun Project

ON A BUDGET?
Created by Icon Solid from the Noun Project

HIP & HAPPENING
Based on a design by Artem Kovyazin from the Noun Project

LOCAL SECRET
Created by hunotika from the Noun Project

WORTH A TRIP
Based on a design by Wahyuntitle from the Noun Project

REGIONAL COOKING
Based on a design by Arthur Shlain from the Noun Project

PERPETUAL WINE
Based on a design by Alex Furgiuele from the Noun Project

KILLER COCKTAILS
Created by Jony from the Noun Project

BEER OH BEER
Created by Paul Tilby

FOREIGN FODDER
Based on a design by mikicon from the Noun Project

ICE CREAM DREAM
Based on a design by Vectors Market from the Noun Project

DELECTABLE DAY TRIP
Created by Iconfactory Team from the Noun Project

SWEET TOOTH
Based on a design by Edward Boatman from the Noun Project

PUB GRUB
Created by Vectors Market from the Noun Project